INSULATION GUIDE FOR BUILDINGS AND INDUSTRIAL PROCESSES

INSULATION GUIDE
FOR BUILDINGS AND
INDUSTRIAL PROCESSES

Edited by L.Y. Hess

NOYES DATA CORPORATION
Park Ridge, New Jersey, U.S.A.
1979

Published in the United States of America by
Noyes Data Corporation
Noyes Building, Park Ridge, New Jersey 07656

FOREWORD

In these energy-conscious times the need for more and better thermal insulation is obvious. Recent rapid rises in fuel costs have confirmed the economic value of insulating as a fuel conserving practice.

Part I of this book deals with "Building Insulation (Residential and Industrial/Commercial)." Thermal insulating materials are assessed and systems for residential and industrial building applications are proposed and elucidated. At the same time that maximum insulation of buildings effects considerable savings in costs and taxes, it confers the added benefit of increasing the availability of fuel, thus providing industry with more fuel to be used for productive purposes.

Part II deals with "Insulation for Industrial Processes (Nonstructural)" and explores the use of thermal insulation by industrial plants and utilities to conserve energy by maintaining adequate process temperatures. Emphasis is on effective temperature ranges vs optimum thickness of insulating materials, their applicability for certain types of processes and the resultant economic rewards, noting the high rate of return on investment from insulating. A chapter on judging heat transmission data should prove most useful and interesting to readers of this book.

Because the information in this book is taken from multiple sources, it is possible that certain portions of this book may disagree or conflict with other parts of the book. This is especially true of monetary values and opinions of future potential. We chose to include these different points of view, however, in order to make the book more valuable to the reader.

Cost figures provided are those given in the report cited, the date of which is always given. When the dates of the cost figures themselves are given, we have included them.

Advanced composition and production methods developed by Noyes Data are employed to bring these durably bound books to you in a minimum of time. Special techniques are used to close the gap between "manuscript" and "completed book." Industrial technology is progressing so rapidly that time-honored, conventional typesetting, binding and shipping methods are no longer suitable. We have by-passed the delays in the conventional book publishing cycle and provide the user with an effective and convenient means of reviewing up-to-date information in depth.

The expanded Table of Contents is organized in such a way as to serve as a subject index and provides easy access to the information contained in this book which is based on various studies produced by and for diverse governmental agencies under grants and contracts. These primary sources are listed at the end of the volume under the heading "Sources Utilized." The titles of additional publications pertaining to topics in this book are found in the text.

Some of the illustrations in this book
may be less clear than could be desired;
however, they are reproduced from the
best material available to us.

CONTENTS AND SUBJECT INDEX

INTRODUCTION

With the current energy shortages threatening the physical comfort as well as the economic welfare of practically every person today, it is apparent that the use of energy conservation methods has become a necessity. As the costs of heating and cooling continue to increase at an alarming rate, one of the more economically attractive methods of conservation is by the use of thermal insulation to limit undesirable heat flow. Thermal insulation is important not only in improving thermal performance of buildings, but, since close to half of the total energy consumption in this country is associated with industrial processes, it likewise is a significant factor in determining the technical and economic success of a process.

Thermal insulation has been contributing to energy conservation and will increase its role in the coming years. As the use of insulation in reducing energy consumption becomes more important to the public, the number of news items, advertisements, government hearings, etc. proliferates; however, consistent, reliable, and factual information about some products has not really been readily available nor in many cases applicable.

A coordinated, concentrated, and cooperative effort by all interested parties—government, building contractors, associations, code organizations, universities, manufacturers, installers, consumer groups—has to be made if progress is to continue. It is necessary to have an adequate assessment of insulation materials and systems in order to apply proper decision-making for both short-term and long-term energy savings.

This book assesses thermal insulation materials and systems for building applications and for industrial processes. It is divided into these two sections accordingly. In the Building Section, various generic insulation materials, their thermal properties, their application in buildings, including retrofitting existing buildings, testing methods, and building codes are discussed. In the Industrial Processes Section, the materials are characterized by their effective temperature range, type of industrial application, and economic effects, including a discussion of optimum industrial thickness. The book concludes with an assessment of the sources of information that are available regarding heat transmission.

PART I

BUILDING INSULATION

(Residential and Industrial/Commercial)

OVERVIEW OF THERMAL PERFORMANCE
OF BUILDING INSULATION

> The material in this chapter was excerpted from a report pre-
> pared by Brookhaven National Laboratory with the assistance
> of Dynatech R/D Company (BNL-50862).

The increasing emphasis being placed upon energy conservation in general and
insulation materials in particular has given rise to a proliferation of conflicting
claims and confusion about the performance of energy saving devices. Accordingly, it is important to have a reliable assessment of thermal insulation materials and systems for building applications for decision making by the federal
government, industry, and consumer. This study may be used to identify areas
where new test methods and standards are needed to establish new programs
for improving thermal performance of buildings, and as a basis for setting new
or improved standards after the recommended test programs have been completed.

Seven major generic types of insulation are identified, based upon the nature of
the material used: mineral fiber (glass, rock, and slag), cellulose, cellular plastics, perlite, vermiculite, reflective surfaces, and insulating concrete. There are
widely varying degrees of concentration of manufacturing capacity within these
categories: for example, there are four manufacturers of fiber glass insulations
and approximately 300 manufacturers of cellulose fiber insulation. Not all products are suited for all building applications, particularly if both new and retrofit
applications are considered.

The manufacturers generally sell insulation products to dealers/wholesalers, to
large insulation contractors or to chain stores. The dealers/wholesalers sell to
retail stores and to insulation contractors. The insulation contractors install
insulation both in new homes for builders or in existing homes for the homeowners. The retail stores and chain stores sell to homeowners and to small contractors. Manufacturers give instructions and recommendations for installation
procedures, but they do not control installation practices by the installers. In
most cases, plant manufacturing quality control is adequate, but problems have
been experienced with some insulation materials processed on-site. Also, there

is evidence that installation of insulation materials is not always well controlled. With the sensitivity to energy conservation, attention to installation details is improving.

The total size of the insulation market is estimated at between $2,000 million and $3,000 million with approximately one-half to two-thirds being for building envelope applications. Fiber glass is the predominant material. Rapid expansion of plant capacity is expected throughout the industry by 1980, e.g., 30 to 40% for mineral fiber products, about 200% for cellulose insulation, and about 900% for urea-based foam insulation.

If expansion plans for each segment of the industry are realized, the relative market share of each material will change, with urea-based foams and cellulose insulation increasing their market shares at the expense of mineral fiber insulation. It has been suggested by others that this expansion will enable the industry to meet the increased demand resulting from programs designed to improve the thermal performance of buildings. The impact of this expansion on raw materials availability is uncertain at the present time.

There are many factors which have been identified as having an effect upon thermal performance of insulations: compaction, thickness, humidity, temperature, and moisture/air barriers. The insulation in turn has an effect on the system in which it is placed due to its fire performance, corrosion, water absorption and similar properties. In general, insulation materials have been sufficiently characterized for properties of generic materials under static (steady-state) conditions, but additional work is necessary to evaluate these materials under dynamic (transient) conditions.

Insulation assemblies for roofs/ceiling, walls, and floors/foundations, typical to residential construction as well as those representative of commercial/industrial applications were reviewed. Data are available based upon static test conditions and calculation procedures are used to estimate performance for average conditions over time. These performance estimates do not take into consideration dynamic effects which may have significant impact on actual field performance. Additional investigation is necessary to determine the effects of dynamic conditions on assembly thermal performance under laboratory and actual field conditions.

The technology for static testing is sufficiently advanced to enable reliable measurement of R-values and other properties. To support this testing there is an urgent need for reference materials. There is also a need for improvement in the technology of dynamic testing which would enable reliable measurement over a wide range of conditions experienced in actual practice. Building codes are now incorporating energy conservation provisions, and the standards reflect the state-of-the-art of testing technology and insulation materials.

Properties of insulation materials are usually presented for the material isolated from the field environment. These data are then used in design calculations to predict the performance of insulation assemblies which are part of the building structure. Various methods employing different assumptions are used for design calculations. This could result in performance predictions which are different from actual performance. There is a lack of sufficient experimental verification of the predicted performance of these systems, both by simulated laboratory experiments and tests in actual field conditions. There is also a lack of sufficient test methods and instrumentation for field evaluation of performance.

Significant gaps do exist, therefore, in the knowledge of properties and performance of insulation materials and systems. Recommendations have been made which should result in better use of thermal insulations for conservation purposes in both new and existing buildings. Some of the important recommendations are summarized as follows:

- Establish a cooperative effort among the various interested parties to obtain the information that is lacking and disseminate existing and new information to those groups which require it.
- Develop standard heat flow reference specimens to be used for calibration of testing equipment.
- Investigate the long-term effects of dynamic conditions (temperature, moisture, weathering, etc.) on the thermal performance and fire resistance of all insulation materials.
- Determine the quantitative relationship between R-value and temperature for all insulation materials.
- Establish quantitative small- and large-scale test methods for fire performance of insulation materials and systems.
- Develop a test method for toxicity of products of combustion from insulation materials.
- Develop a test method for corrosive effects of insulation materials.
- Develop equipment and test procedures for large-scale dynamic thermal performance tests.
- Develop a method for determining the effects of air movement through insulation materials.
- Determine the importance of vapor barriers.
- Develop equipment and procedures for field studies of thermal performance of insulation assemblies.

DESCRIPTION
OF THE INSULATION INDUSTRY

The material in this chapter was excerpted from a report prepared by Brookhaven National Laboratory with the assistance of Dynatech R/D Company (BNL-50862).

HISTORY

Techniques and materials for separating man from the heat and cold in his external environment are as old as civilization itself. Thatched roofs of tropical settlers and adobes developed by American Indians were early versions of insulations. They performed, to some degree, their intended functions, isolating man from the heat of the sun in summer and the cold of winter winds and snow.

Development of insulation materials paralleled that of buildings in which it was installed. It had, until quite recently, one purpose—that of man's comfort. Today, the use of insulation has become more an economic necessity, reducing the loss or gain of heat and the costs associated with that loss or gain.

In the 1800s, insulations used were natural materials, largely vegetable, principally wood and its derivatives. Early in the 20th century, the beginning of what is today the insulation industry, the development and manufacture of materials specifically designed to retard the movement of heat, was started.

The first materials manufactured on a large scale were rock and slag wools. Natural rocks or industrial slags were melted in a furnace, fired with coke, and the molten material was spun into fibers and formed into felts or blankets. These were used in buildings of all types, and also for industrial processes but on a very limited basis where temperatures of an extreme nature existed. The use of these materials gradually increased during the first half of the 20th century and at the end of that period they were being used in most houses as well as industrial/commercial buildings, although still limited in terms of quantities and thicknesses used. In 1928 there were about 8 plants manufacturing these products in the U.S. This number increased to 25 by 1939 and 80 to 90 in the 1950s, but has since declined to about 15 to 20 today.

Glass fibers were successfully developed on a commercial scale in the U.S. in the 1930s by Owens-Illinois, Corning Glass Company, and Owens-Corning Fiberglas Company (a company formed by the first two companies). Fiber glass insulations were developed during the 1940s and 1950s by melting inorganic materials, principally sand, and fiberizing the molten glass into blanket-type material. Owens-Corning Fiberglas was the only producer of fiber glass insulation in the U.S. until 1950. An antitrust action filed in 1949 by the U.S. Department of Justice resulted in a settlement in which Owens-Corning Fiberglas agreed to and did license qualified companies to produce fibers. In addition, alternative processes have since been developed and are now being used to produce glass fiber.

In the mid-1940s, the domestic perlite industry began commercial production. One major application of the crude perlite when processed and expanded was as thermal insulation. At the same time, some entrepreneurs in various parts of the U.S. began grinding waste paper into a fibrous state for use as an insulating material. This was the start of the cellulose insulation industry.

The commercial production of plastic foams began in the mid-1940s with extruded, foamed polystyrene, originally developed by Dow Chemical Company. It was followed in the late 1950s by the commercial production of urethane foams, polystyrene foams manufactured from beads, and urea-formaldehyde foams. Phenolic foams, which have never been very significant commercially in the U.S., have been produced in Europe since the 1950s and in Canada since about 1972. The major emphasis in the first 10 or 15 years of its commercial life was its use in particularly severe insulating applications, such as in low-temperature space facilities, because of its moisture resistance and its highly stable thermal resistance.

Over the last 10 or 15 years, however, the uses for plastic foams, particularly the polystyrenes and polyurethanes, have expanded rapidly into building insulation applications. Plastic foams are often considered to be new as building insulation materials because only in the last 5 years or so they have become widely known in the marketplace.

The 1960s saw rapid development of fiber glass as an insulating material, at the same time as homes and buildings were being better insulated, still largely for comfort purposes. As fiber glass use grew, rock wool (and slag wool) declined. The small rock wool plants, approximately 80 to 90 in number in the 1950s declined to about 15 to 20 today, their product being replaced by fiber glass.

Although patents for cellulosic fiber insulation were issued in the 1800s, the product did not find a firm foundation in the marketplace until late in the 1950s. The primary use of the material was for retrofitting attics, and to a lesser degree, insulation of existing wood-frame sidewalls.

The "Battle of the Fuels" in the late 1950s (electric heating versus fossil fuels) was a major impetus for the insulation industry. Accelerating air conditioning sales found electric utilities with surplus capacity in winter months. The long sought "balanced load" seemed to have its solution in the sale of electric heat.

Aggressive marketing of electric power for heating began a trend to greater insulation usage. The relatively high cost of electric energy was balanced against other heating fuels by the use of full wall insulation, 6" to 8" thick attic insulation, and 2" to 4" thick underfloor insulation. The concept prospered and soon fossil type heating plants were being replaced or displaced with electric heating apparatus and insulation.

About 1970, certain private and government groups began to see energy shortages on the horizon. The insulation that had been used for comfort purposes in the past took on a new role, contributing to more efficient use of a scarce natural resource. Prices of energy started to rise, and the use of insulation became more economically attractive. New homes and buildings were better insulated at the time of construction, and in the mid-1970s the refitting of existing buildings, constructed during times of inexpensive and available energy, began to demand larger and larger amounts of insulation materials.

Today, buildings of all types are being insulated thoroughly; codes and standards increasingly require it. As energy shortages become more apparent and costs for heating and cooling continue to rise, interest in use of insulation for economic purposes continues to grow. In summary, the increased use of insulation is being stimulated by concern for resource conservation, new energy-use provisions of building codes, and simple economics of fuel savings.

TYPES OF INSULATION MATERIALS

In this section a description of the more important insulation materials with a general overview of manufacturing processes for these materials is presented. Principal applications for each of these materials are shown in Tables 2.1 and 2.2.

Table 2.1: Principal Building Applications

Locations	Fiber Glass	Rock Wool	Cellulose	Cellular Plastics	Perlite	Vermiculite	Reflective Surfaces	Insulating Concrete
Industrial*								
New Construction								
Roof/Ceiling	x	—	—	x	x	—	—	x
Walls	x	x	—	x	x	x	—	—
Floors/Foundation	—	—	—	x	—	—	—	—
Retrofit								
Roof/Ceiling	x	—	—	x	x	—	—	x
Walls	x	—	—	x	—	—	—	—
Floors/Foundation	—	—	—	—	—	—	—	—
Commercial								
New Construction								
Roof/Ceiling	x	x	x	x	x	—	—	x
Walls	x	x	—	x	x	x	—	—
Floors/Foundation	x	x	—	x	—	—	—	—
Retrofit								
Roof/Ceiling	x	x	x	x	x	—	—	x
Walls	x	—	x	x	—	—	—	—
Floors/Foundation	—	—	—	—	—	—	—	—

(continued)

Table 2.1: (continued)

Locations	Fiber Glass	Rock Wool	Cellulose	Cellular Plastics	Perlite	Vermiculite	Reflective Surfaces	Insulating Concrete
Residential								
New Construction								
Roof/Ceiling	x	x	x	x	—	—	—	—
Walls	x	x	—	x	—	—	x	—
Floors/Foundation	x	x	—	x	—	—	x	—
Retrofit								
Roof/Ceiling	x	x	x	x	—	x	—	—
Walls	x	x	x	x	—	—	—	—
Floors/Foundation	x	x	—	x	—	—	x	—

*Note: Industrial insulation is described in detail in Part II of this book.

Table 2.2: Market Segments and Product Usage

Location or Building Section	. . . . Products Used	
	New Buildings	Retrofit
Residential Buildings (wood frame)		
Ceilings	fiber glass batt fiber glass loose fill rock wool batt rock wool loose fill cellulose loose fill	fiber glass batt fiber glass loose fill rock wool batt rock wool loose fill cellulose loose fill vermiculite
Sidewalls	fiber glass batt rock wool batt cellular plastic sheathings wood fiber sheathings reflective surfaces	fiber glass loose fill rock wool loose fill cellulose loose fill cellular plastics
Floors	fiber glass batt rock wool batt reflective surfaces	fiber glass batt rock wool batt reflective surfaces
Commercial/Industrial Buildings		
Roofs/Ceilings	cellular plastics perlite board fiber glass boards insulating concrete spray-on cellulose	cellular plastics perlite board fiber glass boards insulating concrete spray-on cellulose
Walls	cellular plastics fiber glass batt perlite vermiculite	cellular plastics fiber glass batt cellulose loose fill
Floors (wood frame)	fiber glass batt rock wool batt	fiber glass batt rock wool batt
Floors (masonry)	cellular plastics	

Source: BNL-50862

Rock or Slag Wool

Rock and slag wool are terms that are used commonly to denote glassy fibrous substances made by melting and fiberizing slags obtained from smelting metal ores including iron ores (slag wool), or by melting and fiberizing naturally occurring rock (rock wool). (Mineral wool is a generic term which includes fiber glass and rock and slag wool.)

In the early manufacture, raw materials, either slag or rock, were melted in a cupola furnace and the molten stream of material was fiberized by passing it in front of a high-pressure steam jet. As the process evolved, more sophisticated designs of the steam jet were made in an effort to obtain more effective fiberization of the molten slag.

The molten stream of slag tends to be broken up into many small droplets by the steam jets, and as these droplets are swept out at high velocity in front of the jet, fibers are formed by the streaming of tails from the heads of the droplets. Not all of the material is converted into fibers by the process, and some nonfiberized droplets, referred to as shot, remain in the product. This unfiberized slag does not contribute to the thermal performance of the product.

Until the early 1940s, almost all rock and slag wool in the U.S. was made with one version or another of the steam jet process. Around the 1940s, however, at least two other processes were developed and commercialized for rock and slag wool production. The Powell process used a group of rotors operating at relatively high centrifugal speeds to collect and distribute the molten stream of slag in a thin film on the surface of the rotors, and then to fiberize the slag by throwing it off the rotors with centrifugal force.

A second fiberization method which came into use at about this time was the Downey process which combined a spinning concave rotor with steam attenuation. The molten stream was distributed in a thin pool over the surface of the dish-shaped rotor, and flowed up and over the edge of the dish where it was caught up in a high-velocity steam or air stream surrounding the dish and fiberized.

Although both the Powell and Downey processes had advantages over the steam jet method, they still produced a product with a relatively high shot content.

Current problems in rock and slag wool manufacture include coke availability, and, more importantly, blending of available slags to get the desired fiber properties. A typical manufacturing process flow scheme for rock or slag wool insulation is presented in Figure 2.1.

Rock and slag wool products, batt materials and loose-fill for blown or poured application, are principally used for residential construction and retrofitting.

Fiber Glass

Fiber glass, in many ways, is similar to rock or slag wool, with refinements due to process evolution and improvements. Fiber glass consists of an inorganic, mineral base and is made into glass wool by several different processes. A process flow chart for the manufacture of fiber glass insulation is shown in Figure 2.2.

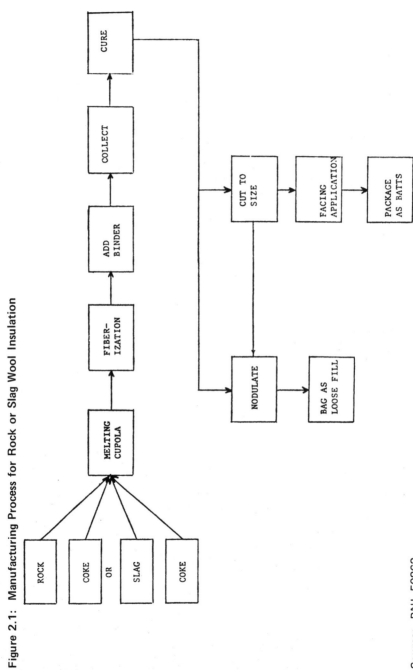

Figure 2.1: Manufacturing Process for Rock or Slag Wool Insulation

Source: BNL-50862

Figure 2.2: Manufacturing Process for Fiber Glass Insulation

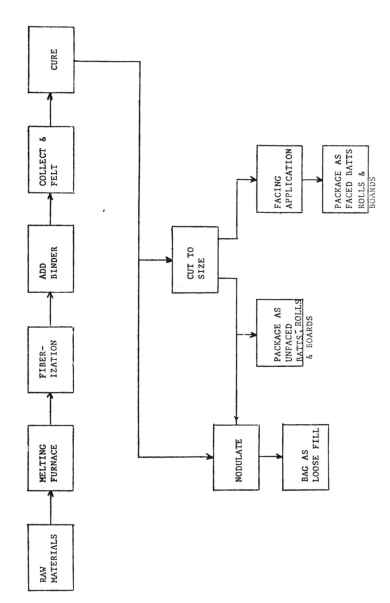

Source: BNL-50862

The glass raw materials are combined and melted in a furnace, and then led out through a forehearth to the fiberization devices. Binder is applied to the fiber as it flows through a collection chamber. Phenolic resin is a binder commonly used by the industry. The fiber with resin is collected on a moving belt and passed through an oven in order to cure or set the resin, and the finished product is taken off the end of the line and packaged.

One type of fiberizing device used with this process uses a rapidly spinning rotor in which the molten glass stream is flowed into the bowl of the rotor, where it is distributed to the sidewall of the spinning rotor. The sidewall contains many small holes, and as the glass flows through these holes it is attenuated by high-velocity jets around the outside periphery of the rotor.

Another process for making glass wool does not utilize the rotor concept, but uses a high-pressure, high-velocity gas jet to fiberize rather coarse primary filaments of glass. In this flame-attenuation process the primary filaments are fed in front of the high-velocity burners, a binder is applied to the attenuated fiber, and it is collected and cured by passing through an oven.

Fiber glass insulation is used in residential and commercial/industrial building envelopes for both new construction and retrofit applications. It is sold as batt, loose fill, and boards, with the usage as indicated in Table 2.2.

Cellulose

Cellulose insulation is made by converting used newsprint, other paper feedstock, or virgin wood to fiber form with incorporation of various chemicals (typically used are borax, boric acid, and aluminum sulfate at a loading of approximately 20% by weight) to provide flame retardancy. This type of insulation is available as loose fill or spray-on with principal applications indicated in Tables 2.1 and 2.2.

Dry Process: The original manufacturing technique for cellulose insulation is a dry process, which is commonly in use today. It consists of feeding the cellulosic material, in most cases newsprint, into one or more mills to shred and pulverize the material into a fibrous, reasonably homogeneous bulk material.

Typical single mill processes operate with up to 150 hp motors. The single mill may function with swinging or fixed hammers or work as a cutter/shredder which forces the final product out through a mill screen with approximately one-quarter inch openings. Paper stock is usually hand fed to allow for sorting of undesirable foreign materials. A dry chemical is usually auger fed directly into the inlet of the mill where it is blended forcefully onto the fibers. The finished product is then air conveyed to a bagging machine for packaging.

This type of operation, which is being utilized by between 75 and 100 plants in the U.S., is not a capital intensive investment. The end product performs as a thermal insulation, but may not meet other criteria set forth by federal and local regulatory bodies, e.g., flammability, corrosiveness, owning to the lack of balanced feed rates for paper and chemicals. Other factors influencing product quality are discussed elsewhere. Production capacity for this type of operation can be as high as 3.5 to 4.0 tons per hour, but is usually around 1.5 to 2.0 tons per hour on a day-in day-out operation.

Multimill processes are designed with many variations to cope with increased output, physical layout, material handling, and quality control problems. They generally follow the basic layout shown in Figure 2.3.

Figure 2.3: Flow Diagram for Multimill Process

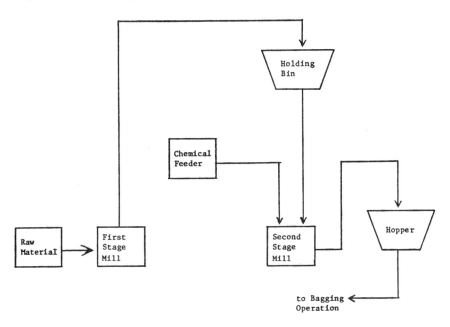

Source: BNL-50862

Paper stock is introduced to the first mill via one or more conveyors. The first mill is usually referred to as a hogger/shredder or cutter. It may perform its task of breaking the paper into approximately three inch to four inch size pieces, sometimes smaller, through the use of fixed or swinging hammers or rotary type cutters.

In some designs shredded paper is passed from the first mill into a holding bin. In others it is immediately fed to a second stage hammer mill where it is reduced into confetti size particles and then fed to a holding bin.

In either process the holding bin is used to achieve a uniform feed rate to the next or last mill in the process. The dry chemical, either a single element or component mix, is usually introduced simultaneously with the ground paper at the second or final mill. This is the most critical stage of the manufacturing process. It is at this point where the human element of physical supervision must be combined with mechanical/electrical controls to maintain a consistent quality in the product.

Introduction of chemicals into the systems may be by pneumatic means (vacuum or forced draft), gravity feed, or auger feed. In many processes the chemicals are preblended, proportioned, and then ground in a mill to the consistency of fine sugar or flour. This finely ground material is dispersed more readily and easily blended into the cellulose fiber materials. The final stage of the process is the bagging operation. The finished product is conveyed into one or more large storage bins which feeds one or more baggers.

Wet Process: The dry process is the more widely used in the industry, but two other methods are available for manufacture of cellulose fiber insulation and can be classified as wet processes. One process involves introduction of an aqueous fire retardant solution sprayed, misted, or sprinkled onto the paper stock material at some point in the process. This is usually done after the paper stock leaves the first mill and prior to entering the finish mill. The method relies on the evaporative ability of the air stream and the short duration heat buildup in the final milling process to remove excess moisture. The amount of chemical, and its evaporative characteristics in response to the process in which it is used, are critical to the properties of the end product. Too little or too much chemical may result in an unmanageable or unusable end product.

The details outlined in the manufacture of dry material are essentially unchanged except for introduction of liquid dispensing equipment to replace bins, augers, chemical mills, etc., used for bulk handling of dry chemicals.

The other wet process incorporates conventional paper making process equipment. It reduces paper stock to a slurry via a pulping process, dewaters it via compression, squeezing up to 50 to 60% of the water from the pulp, and then fluffing and drying the material prior to bagging it as a finished product. There are many potential variations in this process. There are a wider range of opportunities to introduce chemical treatment, dye or other modifications to the process.

There are distinct advantages to either wet process system, but little technical information is available for the evaluation of these processes. Little work has been done in this field due to production economics and as a result, it is being used by very few plants. It is relatively simple and inexpensive to get into the business of manufacturing some form of cellulose fiber insulation by the dry process. The return on investment typically is seldom less than 10% net after taxes. Some operations claim to yield 22% net after taxes. Thus, there is little incentive to invest in untried, yet to be proven wet processes.

Wet processes offer a potentially improved product with better chemical dispersion, fire retardant characteristics, corrosiveness control, and improved resistance to leaching. The wet process may allow for more flexible manufacturing procedures and chemical treatments which are not possible with the dry process. The disadvantage of needing a drying operation with the associated energy requirements should be balanced against the benefits of the process.

Spray-On Cellulosic Fiber Insulation: Application of cellulosic fiber insulation to a given substrate, by means of a bonding agent, has been known for many years. The earliest patent, awarded in the early 1900s, demonstrates a technique for spraying cellulosic fiber and recommends several types of adhesive formulas. Over a period of years, this has evolved into a small segment of the insulation industry with relatively sophisticated application equipment. Cellulosic fiber material for spray-on application is processed in a manner similar to that outlined for dry process material. It will usually have more chemical, generally boric acid, added to it to improve its flame retardant values. Since the product is usually left as an exposed finish, the extra flame retardancy is extremely critical to its in-place performance. Generally, a polyvinyl acetate or an acrylic adhesive is used to bind the material to itself and the substrate. In some techniques, adhesive is mixed with water and sprayed onto the product in fine mist as it is pneumatically conveyed through a special application nozzle. This mist wets the fibers, adds adhesive, and produces the resultant spray-on finish. Another technique has a dry resin added to the product which is activated by a water spray at the nozzle just prior to its installation.

Cellulose Fiber Insulation Flame Retardants: The earliest registered chemical treatment for a flame retardant applied to cellulosic fiber material incorporates the use of boric acid. Since that time many combinations, including or excluding boric acid, have been experimented with in an effort to provide flame retardant properties in cellulosic fiber insulation. These chemicals have been used individually and in combined formulations in an effort to provide minimum cost and acceptable performance as a fire retardant.

Some of the common formulations used contain combinations of the following chemicals: boric acid, borax, aluminum sulfate, lime, ammonium sulfate, ammonium phosphate, mono- and diammonium phosphate, aluminum hydrate, aluminum hydrate, aluminum trihydrate, and zinc chloride. Many combinations have been tried but none have been found totally effective. One combination may produce high resistance to flame spread but permits severe afterglow. Changing the balance of chemicals alters the results, sometimes reversing them. Increasing a component found successful at one concentration may cause failure at another concentration.

Cellular Plastics

There are several different plastics which have use as insulation materials when produced as foams. Both foamed-in-place and board stock foams exist. Due to the significant difference in chemical composition, it is best to discuss separately each cellular plastic insulation.

Polystyrene Foam: Polystyrene, a thermoplastic material, is produced by polymerization of styrene, $C_6H_5 \cdot CH:CH_2$, in the presence of a catalyst. Two methods are widely used for producing polystyrene foam, namely, extruded and expanded.

The extruded form is manufactured by flowing a hot mixture of polystyrene, solvent, and pressurized gas (blowing agent) through a slit into the atmosphere. Due to the reduction in pressure, the gas expands, resulting in a foam with a fine, closed-cell structure. The blowing agent is usually a fluorocarbon.

The expanded (or molded) form is made by placing polystyrene beads (containing a blowing agent) into a mold and exposing them to heat. The vapor pressure of the blowing agent causes expansion of the beads, resulting in a predominantly closed-cell foam. The blowing agent is usually pentane.

Polystyrene foam is primarily used in commercial/industrial buildings as an insulation in ceilings/roofs, walls, and floors/foundations. Use in residential structures includes perimeter slab insulation, foundation insulation, exterior sheathing, and siding backerboard. It has also been used in cathedral type roof/ceiling construction. Application in residential construction is increasing due to more favorable economics as energy costs continue to increase.

Polyurethane and Polyisocyanurate Foams: Polyurethanes are plastics which are the reaction product of isocyanates and alcohols. Polyisocyanurates are made from isocyanates in the presence of a catalyst, resulting in the formation of a more thermally stable isocyanurate ring structure.

Originally, polyurethane foams were formed during the chemical reaction by release of carbon dioxide. However, halocarbons are now used as blowing agents, resulting in an essentially 100% closed-cell foam. Either rigid or flexible foam can be produced, depending on the functionality of isocyanates and alcohols and the molecular weight. For thermal insulations, the rigid foam is used.

Polyurethane and polyisocyanurate foams are produced by several different processes. Continuous slab stock is made by mixing the necessary components and continuously metering the mixture onto a moving conveyor. The mixture forms a continuous foam which can be cut to lengths dependent on ultimate use. Laminates can be made by a similar process, dispensing the mixture between sheets.

Foamed-in-place polyurethanes and isocyanurates are prepared by mixing or metering the components and dispensing them either manually or automatically. Specially designed units are available for spray-on applications.

Rigid polyurethane and polyisocyanurate foams are used primarily for commercial/industrial buildings as roof insulation, floor and foundation insulation, cavity wall insulation, and interior and exterior spray-on wall insulation. Both board stock and foam-in-place forms are available. There is increasing use of these foams in residential construction, principally as sheathing.

Urea-Formaldehyde and Urea-Based Foam: The major application of urea-formaldehyde and urea-based foam as a thermal insulation is for retrofitting residential wall cavities. It is also used, but to a lesser extent, as an insulation in commercial/industrial buildings for both new construction and retrofit and in residential new construction.

Urea-formaldehyde foam is generated at the site of application by the combination of an aqueous solution of a urea-formaldehyde based resin, an aqueous solution foaming agent which includes a surfactant and acid catalyst (or hardening agent), and air. In the mixing or foaming gun compressed air is mixed with the foaming agent to produce small bubbles which are expanded and coated with the urea-formaldehyde resin. The foam is delivered at about 75% water by weight and immediately begins to cure. The foam typically cures sufficiently to be self-supporting in less than one minute while full chemical curing requires several weeks. The rate of water-dry-out from the foam depends on the type of structure in which the foam is applied. The foam produced in this manner consists of an approximately 60% closed-cell structure.

Urea-based foam insulation has been considered a generic material. However, it is known that differences exist in the composition and properties of the foams available in the U.S. The specific formulations used by the industry are proprietary and include chemical components for the purpose of improving foam properties. Also, different types of apparatus are used to produce the foam.

Perlite and Perlite Mineral Board

Perlite is a naturally occurring, siliceous, volcanic glass containing between two and five percent water by weight. Perlite ore is composed primarily of aluminum silicate. Crushed ore particles are expanded to between 4 and 20 times their original volume by rapid heating to a temperature of 1000°C, which vaporizes the occluded water and forms vapor cells in the heat-softened glass.

Perlite is used primarily in industrial/commercial buildings as a roof insulation board material. The next largest use is in lightweight insulating concrete where expanded perlite is mixed with Portland cement. A wide range of density is available. Perlite insulating concrete, both preformed and cast-in-place, is used primarily for roof decks, floor slabs, and wall systems. Low density expanded perlite is used as a loose fill insulation.

Vermiculite

Vermiculite is a mica-like hydrated laminar mineral consisting of aluminum-iron-magnesium silicates with both free and bound water. When the mineral is subjected to high temperatures it expands due to formation of steam which is driven off, thereby causing the laminae to separate. A wide range of density is available, depending upon the amount of expansion.

The principal thermal insulation applications for vermiculite are for residential attic loose fill and as an aggregate in lightweight insulating concrete for roof decks.

Insulating Concrete

Insulating concrete, made with perlite or vermiculite aggregate, has already been mentioned. Insulating concrete can also be made by foaming a Portland cement mixture. When a foaming agent, such as aluminum dust, is added, a chemical reaction occurs evolving a gas, which produces a closed-cell porous material. A wide range of densities is possible, density being a function of the amount of gas evolved. The concrete normally will set in about 3 days to sufficient strength to permit roofing.

Insulating concrete is primarily used for roof systems and decks. Higher density concretes can also be used for wall insulation, where the increased strength is balanced against poorer insulating properties for the higher density material.

Reflective Surfaces

Aluminum foil, the major reflective surface insulation, is different from other insulating materials by the manner in which it retards transfer of heat, namely by reflection of incident infrared radiation. This reduces radiant heat transfer.

Aluminum foil facings are used in conjunction with mineral fiber and rigid board plastic foam insulations and will add to thermal performance where a dead air space is adjacent to the foil. Multilayer aluminum foil is used to insulate residential building walls and floors.

Insulation Board

Board insulation is used in commercial/industrial buildings as part of the roof/ceiling system. For flat roofs, insulation board is placed on the roof deck and the built-up roofing is applied directly to the board. There are several different types of insulation board, including perlite board, foamed plastic board, fiber glass board, laminated board, consisting of perlite or glass and foamed plastic, wood fiber board, and sandwich panels.

MANUFACTURING FACILITIES

Information concerning the location and number of insulation manufacturing plants has been developed. The locations of insulation manufacturing facilities are given in Figures 2.4, 2.5 and 2.6, for mineral wool (including fiber glass and rock and slag wools), cellulose and plastic foam, respectively. After location of these plants was determined, an analysis was made to ascertain what market influences based on population census exist, if any, which led to these manufacturing locations.

Several trends have been noted as a result of this analysis. In general, insulation materials are high bulk products which tend to have relatively high transportation costs. Hence, markets are a primary determinant of plant location. Raw materials, which tend to be relatively dense, have a lesser influence on plant locations. Some processes require larger plant sizes to be economical; hence, these plants are fewer in number.

Manufacturers of rock and slag wool are influenced primarily by proximity to raw material supply as well as their access to the largest markets. The same holds true to some extent for manufacturers of fiber glass products.

The plastic foam insulation manufacturers appear to locate closer to their market with less dependence on raw material supplies. Generally, for the large manufacturing plants more research goes into selecting the best location for plant site with all the normal trade factors considered prior to making a final decision on location.

The cellulosic fiber manufacturers apparently locate close to their market, although there is some dependence on paper supplies. There is no dependence on the local availability of flame retardant chemicals, such as borax and boric acid. These chemicals are routinely shipped from their West Coast source to East Coast consumers. The basic requirements for locating a cellulose fiber insulation plant appear to be dependent on the interest of individual investors as well as availability of a receptive market.

Cellulose fiber manufacturers at one time shipped their products distances of up to 600 miles via truck, but now find it economically unfeasible to do so and seldom ship more than 100 to 150 miles from a plant. Other manufacturers of insulation materials, especially with products of a denser nature, may still find it economical to ship distances of 500 to 600 miles. These shipments may be by railcar or truck as appropriate to the specific situation.

Figure 2.4: Mineral Wool Plants

Source: BNL-50862

Figure 2.5:   Manufacturers of Cellulosic Fiber Products

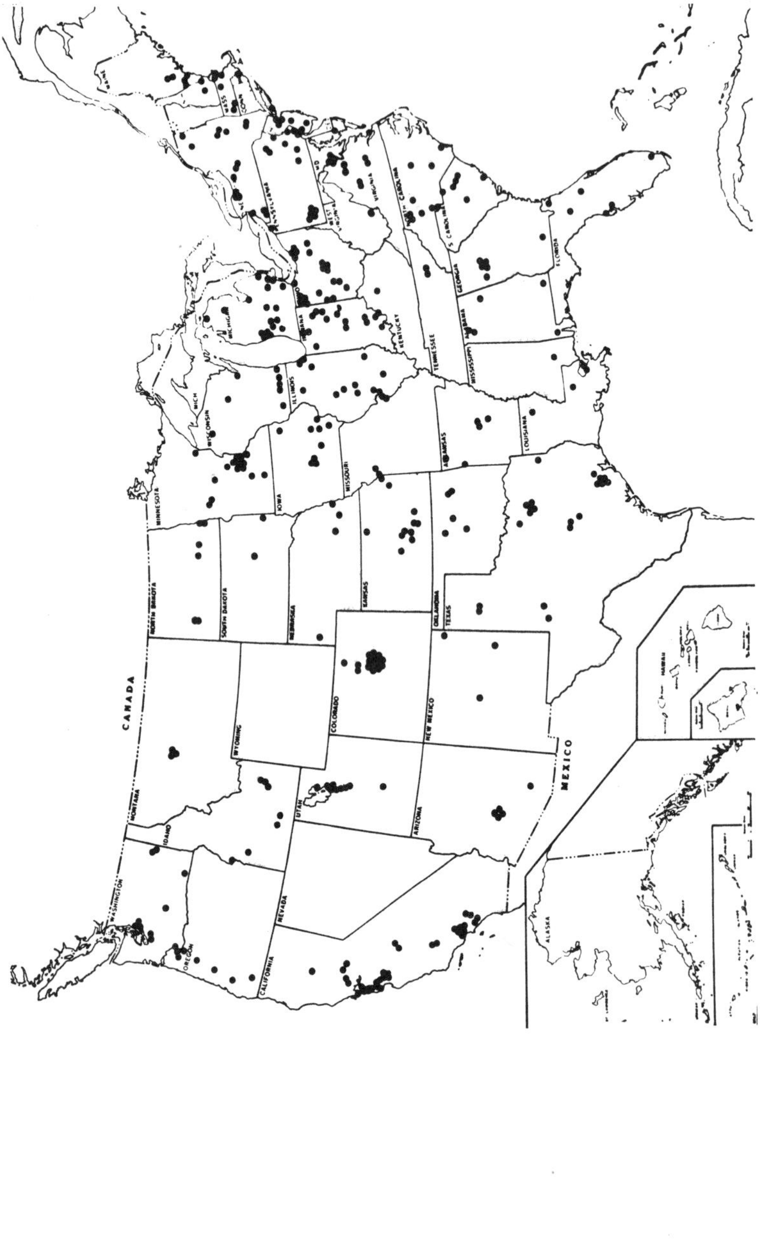

Source: BNL-50862

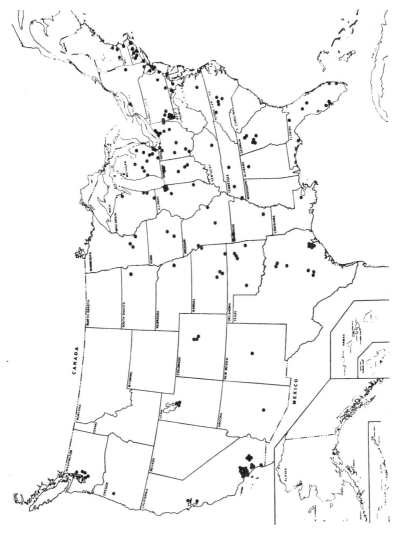

Figure 2.6: Plastic Foam Manufacturers

Source: BNL-50862

INSULATION MARKET

The size of the building envelope insulation market is extremely difficult to assess. Different products report by different category definitions, and numbers which are reported could include equipment, pipe, and duct insulations and products for acoustical purposes in addition to building envelope insulation.

In order to obtain an estimate of market sizes, several different sources of data were used. Questionnaires were sent to insulation manufacturers. A survey was made by the Office of Business Research and Analysis (BRA), U.S. Department of Commerce and other reports such as by ICF, Inc. and by the Council on Wage and Price Stability, have also been used as data sources.

The overall size of the insulation market has been reported to be between $2,000 million and $3,000 million with approximately one-half to two-thirds being for building envelope insulation applications. The market size for each of the various types of insulation materials is discussed below.

Rock or Slag Wool

The number of rock and slag wool manufacturers has been indicated to be between eight and ten (1)(2)(3).

The rock or slag wool industry capacity for structural insulation as of January 1, 1977 was reported to be 917 MM ft^2 of R-11 equivalent (batts and blankets) and 491 MM ft^2 of R-19 equivalent (loose-fill) (1). These numbers were obtained from the manufacturers, who also projected industry capacity as of January 1, 1980 to be 30% greater for batts and blankets and 75% greater for loose-fill (1). The numbers in the OBRA report include some building envelope insulation for industrial use. Residential building applications account for about 90% of this total.

This predicted rise in the industry capacity differs from the trend of approximately a 2 to 3% annual increase since 1960. The sharp upturn predicted by producers is probably due to an expected increase in insulation demand as a result of increasing energy costs.

Fiber Glass

Fiber glass is produced by four manufacturers, Owens-Corning Fiberglas Corp., Johns-Manville Corp., Certain Teed Corp., and Gebr. Knaut Westdeutsche Gipswerke. The OBRA report (1) indicates that as of January 1, 1977 production capabilities for building envelope insulation were 8 billion ft^2 of R-11 equivalent (batts and blankets) and one-half billion ft^2 of R-19 equivalent loose-fill. The predicted expansion as of January 1, 1980 is 35% for the R-11 equivalent and 53% for R-19. The OBRA report values, converted to a weight basis, give a capacity of about 1.56 billion pounds. This compares with total fiber glass structural insulation shipments of 1.38 billion pounds as reported in the 1976 Department of Commerce, Bureau of Census, "Current Industrial Reports." The Council on Wage and Price Stability Report (4) indicates an estimated capacity of 2.3 to 2.5 billion pounds.

The growth rate for the fiber glass insulation market has averaged about 12% annually since 1961, which is about the same as the growth rate forecast by the manufacturers in the OBRA report. Over the last five years, manufacturers have doubled their capacity in anticipation of increased demand (5). Other indications are that the industry expects to expand capacity by about 40% by 1980 (6), a continuation of present growth trends.

Cellulose Fiber Insulation

The size of the cellulose fiber insulation industry is extremely difficult to assess because there are many small manufacturers and the number of these facilities is changing every week. An estimated 70 to 100 producers began operations in 1977 (6). The OBRA report (1) estimates approximately 250 manufacturers as of June, 1977 and the ICF report (2) indicates between 200 to 300 producers. It was estimated that there were about 400 plants at the end of 1977. The production capacity was reported as 1.7 billion pounds (1). Data presented in the ICF report (2), 0.6 billion lb/yr, are significantly lower than the OBRA value. This study indicated production rate for over 300 manufacturers to be about 10% less than the OBRA report value.

The threefold expansion indicated in the OBRA report is in agreement with the growth rate over the last few years, but is greater than the twofold increase indicated in the ICF report. However, certain industry factors should be mentioned which could affect the future rate of production, specifically raw material supplies, including newsprint and chemicals for flame retardants.

There have been reports of a boric acid shortage (7), boric acid being the major component of flame retardant chemicals for cellulose fiber insulation. Boric acid is used in a ratio of approximately 3 pounds boric acid per 40 pounds of newsprint so that the annual demand of boric acid for cellulose insulation is about 105 million pounds, and with expected increase in cellulose output, the boric acid demand could increase to over 100 million pounds by 1980. The 1977 boric acid capacity is about 400 million pounds per year, with U.S. Borax and Chemical accounting for about 65% of this amount. U.S. Borax plans a 30% expansion, expected to be completed by 1980. Other increases in availability could result from imports, particularly from Turkey, Russia and South America.

Even with the increase in boric acid supply, there still could be future problems with availability due to the much higher rate of increase in demand predicted for cellulose insulation. Developments of new flame retardant formulations requiring less boric acid could alleviate potential supply problems. Several chemical companies, including U.S. Borax and Chemical (7)(8) are now in the process of investigating different formulations requiring less boric acid.

The ICF report indicates that, except for spot shortages, waste paper for cellulose insulation is available. However, as indicated above, an increasing demand for this raw material supply is expected which could result in higher costs. This could have a chilling effect on new entries into the market, and result in less than predicted future capacity. This, in fact, could account for some of the difference between the OBRA and ICF forecasted relative increases in production capacity.

Cellular Plastics

The consumption of plastic foams for building insulation has been growing at a rate between 10 to 15% annually. This consists primarily of polyurethane, polyisocyanurate, polystyrene and urea-formaldehyde foams. The growth rate of the plastic foam has been estimated to be 11% annually through 1985 (8)(9)(10). This growth rate thus extrapolates from 400 million pounds in 1974 to a production of about 700 million pounds for 1977.

Polyurethane and Polyisocyanurate Foams: The OBRA report gives the polyurethane and polyisocyanurate foam production capacity as 528 million board feet for foam board and 447 million cubic feet for foam-in-place applications (1). However, this latter amount differs drastically from other sources of information. The difference is most probably due to the use of cubic feet rather than board feet in the OBRA report.

An article in *Rubber and Plastics News,* May 17, 1976 (11), indicated on-site application capacity for 1976 was 375 million board feet and foam board capacity of 407 million board feet. *Plastics World,* August, 1975 (12) indicated 291 million board feet for in-place and 520 million board feet for board stock. On a weight basis, the 1976 output was about 200 million pounds. The same total production (in pounds and board feet) was given in *Chemical Marketing Reporter,* May 10, 1976 (13).

The indicated growth rate for polyurethane foam is about 10 to 15% annually, so that extrapolation to 1977 gives a capacity which is equivalent to that in the OBRA report. Polyurethane/polyisocyanurate foams are primarily used for commercial/industrial applications. Wider use for residential construction is expected in the future, particularly for wall insulation to obtain a resistance of R-19 (14), which could be the standard minimum insulation R-value established in the future. There are approximately 35 manufacturers of polyurethane and polyisocyanurate foams, about 10 of which produce boardstock.

Polystyrene Foams: Polystyrene production capacity was approximately one billion board feet in 1977 (1) with a 32% expansion anticipated by 1980. This is in agreement with information presented in the *Foamed Plastics Directory* (15), namely 75 million pounds in 1976 with expansion to about 98 million by 1980. The only producer of extruded polystyrene foam board is Dow Chemical Co. There are approximately 100 producers of expanded bead polystyrene foam.

Urea-Formaldehyde Foam: There are four major producers of the chemicals for urea-formaldehyde foam, namely, Borden Chemical, Brekke Enterprises, C.P. Chemical Company and Rapperswill Corporation (16). The indicated capacity in the OBRA report (1) is 158 million pounds with Rapperswill having about 80% of the U.S. market (17).

The forecast capacity increase by 1980 is to 1549 million pounds, about a tenfold increase. U-F foam is used primarily for residential retrofit applications. It is estimated that demand for this application will be met by 1983 (1). This could result in a lesser expansion than predicted due to the unwillingness of manufacturers to invest in such a short term market.

Perlite

Expanded perlite is produced by over 30 companies in the U.S. The production for loose-fill thermal insulation has increased from 14 million pounds in 1964 to 24 million pounds in 1968 and 32 million pounds in 1973 (18), an historic growth rate of 9% annually. The 1977 production capacity for loose-fill perlite presented in the OBRA report (1) is 66 million cubic feet which, assuming a density of 6 lb/ft^3, is approximately 400 million pounds per year. It appears that firms may have reported total capacity for expanded perlite without regard to end use, since this is a tenfold increase over historic growth trends for loose-fill perlite insulation.

The primary use of perlite is in aggregates, such as perlite concrete, plaster and insulating board. The consumption of perlite for these applications, much greater than for loose-fill, was about 500 million pounds in 1964 and 600 million pounds in 1973, with a forecast of 1.5 billion pounds by the year 2000 (18)(19).

Vermiculite

The production capacity of vermiculite for loose-fill thermal insulation was 116 million cubic feet in 1977 (1), which is approximately 500 million pounds. This is about a 9% annual increase over the 225 million pounds consumption in 1968 (20). The 119 million cubic feet forecast for 1980 in the OBRA report (1) is an indication that vermiculite use is not expected to increase significantly.

Reflective Surfaces

The production capacity for 1977 for reflective surfaces, mainly aluminum foil, is given in the OBRA report as 854 million square feet with minimal expansion expected by 1980.

REFERENCES

(1) Survey Report: "U.S. Residential Insulation Industry," Office of Business Research and Analysis, U.S. Department of Commerce, (August, 1977).
(2) "Supply Response to Residential Insulation Retrofit Demand," ICF Inc., report submitted to The Federal Energy Administration, Contract P-14-77-5438-0, (June 17, 1977).
(3) National Mineral Wool Insulation Association, (October 18, 1977).
(4) Council on Wage and Price Stability, Executive Office of the President, report on the Fiber Glass Insulation Industry, (June 14, 1977).
(5) Simison, R.L., *The Wall Street Journal,* page 48, (September 27, 1977).
(6) *Business Week,* page 88, (September 26, 1977).
(7) *Chemical Week,* page 29, (June 29, 1977).
(8) *Chemical Marketing Reporter,* page 7, (April 10, 1972).
(9) *Chemical Engineering,* page 46, (August 16, 1976).
(10) *Chemical Marketing Reporter,* page 77, (December 13, 1976).
(11) *Rubber and Plastics News,* page 10, (May 17, 1976).
(12) *Plastics World,* page 9, (August 18, 1975).
(13) *Chemical Marketing Reporter,* page 7, (May 10, 1976).
(14) *Modern Plastics,* page 44, (October, 1977).
(15) *U.S. Foamed Plastics Markets and Directory,* Technomic Publishing Co., Westport, CT (1977).

(16) Rossiter, W.J., Jr., Mathey, R.G., Burch, D.M. and Pierce, E.T., NBS Technical Note
        946, (July, 1977).
(17) *Energy Users Report,* No. 210, page 16, (August 18, 1977).
(18) *Mineral Facts and Problems,* page 781, (1974).
(19) *Bureau of Mines,* 650, page 1135.
(20) Ibid, page 1288.

ASSESSMENT
OF THERMAL INSULATION MATERIALS

The material in this chapter was excerpted from a report prepared by Brookhaven National Laboratory with the assistance
of Dynatech R/D Company (BNL-50862).

INTRODUCTION

In undertaking an assessment of the properties of the current insulation materials
it is first necessary to define the parameters for which properties information is
required. Following this it is then necessary to try to obtain the information on
which an objective study can be made.

In the majority of cases manufacturers of the products do provide data sheets
containing some basic information on the properties. In general, information is
reported on thermal performance, fire performance, water vapor transmission,
density and mechanical properties where relevant to a material. However, comprehensive information is not available for all products.

An examination of this information shows it often provides limited data, with
other properties inferred by reference to a specification. There is no uniform
method utilized by manufacturers to present material property data. For a few
generic materials, values of one or more properties, particularly thermal performance, can differ very significantly.

The thermal performance of a material (or system) is usually quantified in terms
of thermal resistance (R-value) which describes the ability of that particular material (or system) to restrict the flow of heat.

Thermal resistance is defined as a property of a particular material or material
assembly measured by the ratio of the differences between the average temperature of two surfaces to the steady-state heat flux in common through them and
is expressed in units of hr-ft^2-$^\circ$F/Btu.

Most of the information has been generated by the manufacturers as part of their research, development, and quality assurance programs. Much of this information is obviously proprietary and, therefore, not available in published literature.

One source of "design values" is the ASHRAE Book of Fundamentals (Chapters 18 and 20 in editions up to 1972 and Chapters 20, 22 and 25 in the 1977 edition). Design values have been agreed upon within the relevant TC 4.4 Thermal Insulation and Vapor Barrier Subcommittee and based upon three independent sources of information submitted to the committee. The majority of the information is for thermal performance with some additional data on water vapor transmission characteristics.

There was an independent study on cellulose fiber material (1), but it did not cover all of the relevant properties. The results of a comprehensive study on urea-formaldehyde were published in Reference (2). For polyurethanes and polystyrenes in particular, some papers containing information concerning mechanical properties, thermal performance, effects of aging on thermal properties together with other related properties have been published (3).

In considering the properties of thermal insulating materials which are important to overall satisfactory performance in buildings, a number of parameters are important. Of these, the thermal performance predominates since the application is for thermal insulation. Resistance to fire and the production of toxic gases during combustion are factors, as are water vapor permeability and absorption. However, while accurate laboratory-performed tests can provide information on all of these properties, the results may not adequately represent the performance of the materials in use or after installation for a period of time.

Little public data are available on the effects which parameters such as aging, temperature cycling, effects of moisture, etc., may have on the overall performance of these materials.

An attempt was made to establish whether additional information did exist and to obtain it for the assessment, by means of a questionnaire sent to manufacturers of all products being considered. While trying to obtain confirmatory information for some of the basic properties which were already available, the questionnaire requested information on other factors, such as air permeability, aging and degradation effects.

The overall assessment of the properties was based, therefore, upon information obtained from these four major sources: manufacturers' literature; open literature; results of questionnaires; and knowledge of experts in the insulation field.

However, in all cases the information for a particular generic material or product was reviewed in relation to some of the major factors which can influence the properties, particularly thermal performance. In this way the individual properties subsequently reported are consistent with basic physical principles and thus can be accepted as a reliable guide for present use. Some of these basic considerations are now outlined prior to the review of the individual materials.

THERMAL PROPERTIES

This area is the one where a large number of factors have to be considered in relation to the overall applicability and validity of information available. The energy crisis has resulted in a need to reassess energy conservation practices. The approach to energy conservation, proposed by the insulation manufacturers long before the energy crisis, was to invest in insulation to reduce equipment and operating costs. Insulation manufacturers have recommended the practice of economic thickness, particularly in industrial applications (4).

Described here are a number of basic concepts related to the use of the insulation and the general practice of energy conservation which, in the end, will result in optimizing the return from the insulation investment. Today, there is extensive knowledge of the principles of heat flow and the performance of insulation, but little quantitative information is available on the possible detrimental effects introduced by various construction practices.

Before proceeding with this aspect, the basic heat transmission mechanisms in an insulation are outlined in order to appreciate how the insulation retards the flow of heat.

Mechanisms of Heat Transfer

Radiation is the primary mode of heat transfer between two surfaces at different temperatures. Conduction and convection are important only as they interfere with the primary mode, radiation. The objective of the insulation in relation to these modes is to:

> minimize radiation transfer;
> minimize convective transfer;
> introduce a minimum of solid conduction; and
> minimize gas conduction.

To obstruct the radiative transfer, as many radiation absorbers as possible are placed between the temperature boundaries without significantly increasing the solid conduction. Most building insulations consist of many small voids, containing air, separated by a small fraction of solid material. The solid contributes only minimally to conduction while the small voids inhibit convective transfer. For materials containing air, which has a thermal conductivity of approximately 0.18 Btu-in/hr-ft^2-°F at 75°F, the maximum R-value per inch must be below 5.6 hr-ft^2-°F/Btu. However, some plastic foams, which are closed cell, contain a fluorocarbon gas which has a much lower thermal conductivity than air and, therefore, can yield a higher R-value per inch.

Heat Transfer Through Thermal Insulations

To examine the mechanisms of heat transfer within the insulation, the present familiar term "thermal conductivity" or "k" will be used. This is the simplified expression describing the combined conductive effects within a very complex system.

The term itself (k) and its definition, referring to an intrinsic property of a homogeneous material, is subject to theoretical questioning when referring to insulations and this point is discussed later. However, in many instances when using insulation, it is still a practical way to define the characteristics of an insulation.

The apparent thermal conductivity resulting from the parallel mechanisms of heat transfer may be defined as

(1) $k_{app} = k_g + k_{cv} + k_s + k_{rc} + k_{rt} + k_i$

k_g = k of air or other gas x (1 – f) for 0.01 $<$f$<$0.25; where f is the volume fraction of the solid portion of the system

k_{cv} = k of convection under specific conditons

k_s = k of solid particle to particle conduction

k_{rc} = k of radiation conduction (inter-particle radiation)

k_{rt} = k of radiation transmission (influenced by bounding surfaces)

k_i = k attributable to interaction with gas

k_g: The thermal conductivity of air or other gas in low density materials is nearly a constant since the major portion of the structure consists of the gas (for example, 99% air volume in 1.6 lb/ft^3 mineral fiber and not less than 75% for most insulations).

k_s: The thermal conductivity of the solid is related to the number of discontinuities in the network (contact between the fibers or other solid particles). It should be kept to a very small portion. It is also related to the gradient along the fiber or structural component, which is influenced by the orientation of the individual fiber. Thus, many insulations are anisotropic in the thermal conduction properties.

k_{cv}: The natural convective transfer appears only when the cells are large enough to allow natural convective air turbulence. These cells usually must be several millimeters across for the phenomenon to become significant. For example, at room temperature, in a 0.50 lb/ft^3 density mineral fiber insulation with a uniform 4 micrometer (micron) fiber diameter distribution, the average pore size is approximately 1 millimeter. However, because of nonuniformity of the fiber distribution, a small amount of convection will occur in slightly higher densities (0.60 lb/ft^3 approximately with 4.5 micrometer fiber).

k_r: Radiation is a significant portion of the total heat transfer at room temperature in insulations having densities lower than 2 lb/ft^3. At higher densities and elevated temperatures, it is also significant. This is the portion which influences the shape of the k_{app} versus density, k_{app} versus temperature, and k_{app} versus thickness curves of any insulation. The radiation contribution consists of two parts: (1) k_{rt}, the radiation transmission is the portion of the overall radiant heat transfer not absorbed, but rather reflected and scattered by the fibrous structure; (2) k_{rc}, the fiber-to-fiber reradiation, or radiation conductivity, which is dependent upon the absorptance and emittance of the fiber and the temperature difference between fibers.

k_i: The interaction between the air and the solid components of the system. The air conduction acts as a short circuiting process since it contacts all radiation absorbing surfaces.

In order to appreciate the relative level of these terms and how they contribute to a given thermal conductivity value, the thermal conductivity of a typical glass fiber insulation is taken as an example (5)(6). Figure 3.1 shows these mechanisms under very specific conditions and Table 3.1 shows how each of these mechanisms is modified by the presence of glass fiber insulation or modification of the boundaries (6). The mechanics of heat transfer illustrated in this example can help explain the mechanisms as they operate in other insulating materials used as building insulations.

Figure 3.1: Thermal Conductivity Components Versus Density for a Typical Glass Fiber Insulation (6)

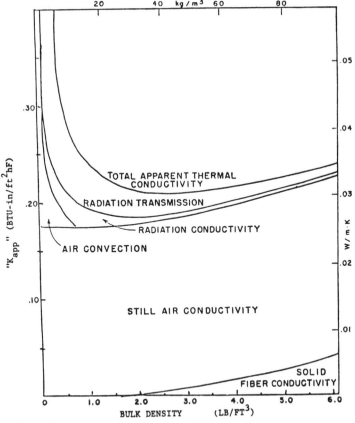

[100°F (38°C) hot face, 50°C cold face, 1.0" (25.4 mm) thickness, black boundaries]

Source: BNL-50862

Table 3.1: Effect of Glass Fiber Insulation on Mechanisms of Heat Transfer

| Source of Heat Flow | Black Surface, ϵ = 0.95 | | Gold Surface, ϵ = 0.08 | |
	No Insulation	0.66 pcf Glass Fiber	No Insulation	0.66 pcf Glass Fiber
	. . . . (Btu-in/hr-ft^2-$^\circ$F)			
Radiation	1.054	0.077	0.046	0.030
Conduction				
Air	0.178	0.177	0.178	0.177
Solid	0	0.016	0	0.016
Convection	0.435	0	0.435	0
Interaction	–	0.014	–	0.047
Total K$_{app}$	1.667	0.284	0.659	0.270

Note: Conditions of tests—upward heat flow, hot face 100°F, cold face 50°F.

Source: BNL-50862

An insulation material will perform as designed if all factors remain as initially assumed for the design. Actual performance could be affected by such factors as packaging, handling, installation, and the environment.

Insulations Used as Building Insulations

Figure 3.2 shows the k versus density relationship of several materials commonly used as building insulations. The basic mechanisms of heat flow can explain the relationships of these varied materials. However, there are differences between materials and these differences can be of significance in selecting the materials for an application.

Factors Affecting Insulation Performance

The complexity of the mechanics of heat losses or gains through building envelopes is such that no single solution will estimate accurately the individual values or average savings which may be realized on different installations under varying conditions. The thermal performance of an insulation material is one factor in the prediction of energy savings. It is intended to be used as input for calculations to design and estimate the thermal performance of a building envelope. The design methods should also recognize where individual application factors may produce significant deviations from the average design estimates.

Factors which appear important in determining how well insulation will perform in place are:

(1)   The actual R-value of the insulation as found in the market place.

(2)   The actual thickness as installed.

(3)   The quality of installation (open spaces, holes, vapor barrier, or better yet, air barrier to eliminate infiltration, etc.).

(4)   Changes resulting from time, vibration, moisture, air movement, etc.

A limited study was conducted by the Minnesota Energy Agency for the Department of Energy, of some of these effects with different insulation materials in existing residential buildings.

Figure 3.2: Apparent Thermal Conductivity Versus Density of Various Thermal Insulations Used as Building Insulations

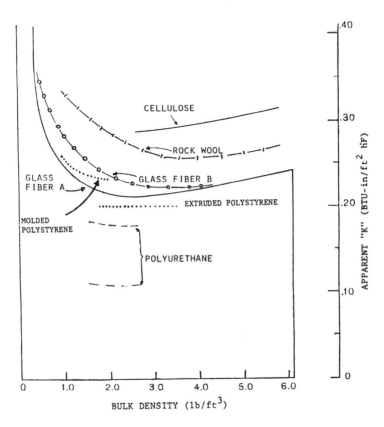

Source: BNL-50862

Technology does not exist to measure adequately the in-place thermal performance of thermal insulation in a building system. A number of field measurement devices have been proposed, but none has been proven to give quantifiable results. There is a need to provide technology which will enable the user to ascertain that insulation being purchased or having been installed really provides the published performance values. It appears that in a number of cases the anticipated thermal resistance and related energy conservation are not being achieved. Individual effects which can alter thermal performance are described below.

Compaction Effects: Loose-Fill or Blown Insulation — With all loose-fill or blown insulations it is possible to apply the insulation lighter than the recommended coverage directions given on bag labels, thereby achieving a greater coverage. With time, the product so applied, tends to densify by settling with a consequent reduction in thickness resulting in an overall loss in thermal resistance. However, when installed in accordance with manufacturer's recommended coverage directions, settling should still result in a product which meets design specifications.

This factor is illustrated by calculations presented in Tables 3.2 and 3.3 for typical blown cellulose fiber and fiber glass respectively. It should be noted that for the glass fiber product, the loss of thickness has a negative impact on resistance but this is somewhat offset by an improved k_{app}. Loss of thickness for cellulose insulation has a negative impact on both resistance and k_{app}.

Table 3.2: Cellulose Insulation—Calculated Effect of Compaction on Installed R

Compaction (%)	Thickness (in)	Density (lb/ft^3)	K_{app} (Btu-in/hr-ft^2-$^\circ$F)	Installed R-Value	Reduction in R-Value (%)
0	5.5	2.5	0.274	20.1	—
10	5.0	2.75	0.277	18.1	10
20	4.5	3.06	0.280	16.1	20

Source: BNL-50862

Table 3.3: Loose-Fill Fiber Glass Insulation—Calculated Effect of Compaction on Installed R

Compaction (%)	Thickness (in)	Density (lb/ft^3)	K_{app} (Btu-in/hr-ft^2-$^\circ$F)	Installed R-Value	Reduction in R-Value (%)
0	6.0	0.82	0.300	20.0	—
10	5.4	0.91	0.290	18.6	7
20	4.8	1.03	0.277	17.3	13

Source: BNL-50862

Preformed Insulation Blankets, Rolls and Batts — Blanket-type insulations must have proper recovery of thickness after packaging in order to perform as specified. Insulations of this type, by nature, are bulky and in this form they are normally packaged with a compression of 6 or 7 to 1. The resiliency of the bonded fibers should allow the product to recover at least to the design thickness to ensure that the claimed thermal resistance will be met.

An experimental study was undertaken by Dynatech, on behalf of the U.S. Department of Energy, to investigate the validity of the various parameters, including thickness recovery, which affect thermal performance of all currently available R11 and R19 mineral fiber batt products. The results of the study are not yet available.

Moisture — The need of vapor barriers to stop or minimize the flow of moisture into the insulating components was demonstrated in the early days of insulation usage in cavity walls. The problem was recognized when insulation installed in cavity walls resulted in paint blistering on the outside surface of the building.

The higher level of moisture inside the dwelling caused a migration to the outside where the moisture content was lower. The solution to the problem involves the use of a good vapor barrier on the inside surface and/or making sure that the permeance to moisture of the outside surface is greater than that of the inside surface. Under these conditions there are relatively few cases where moisture would accumulate to great enough levels in the insulation to destroy its overall effectiveness.

The requirement that a vapor barrier be applied to the insulation is an attempt to provide a good moisture seal on the inside surface; unfortunately, the vapor barrier is perforated by electrical outlets and other service connections in the wall. The most ideal seal is obtained by using a plastic film under the interior of the gypsum or wall finishing material.

In most instances, moisture vapor movement under these conditions will cause very little detrimental effect on open cell insulation performance unless there is real accumulation (as a result of water leaks) of water high enough to saturate the insulation. There may be more serious damage to the building components which will hold and absorb this water.

Suggestions have been made that moisture vapor may have some very damaging effects on the performance of some insulations, particularly fibrous materials, where it is supposed that the fiber will tend to hold water, slump and form a dense mass with poorer insulating properties. While no controlled laboratory or field test results are available to prove or disprove this suggestion, thousands of homes which have been insulated with fibrous insulation do not appear to show any detrimental effects of this nature.

Studies at the National Bureau of Standards (7) have indicated that the air movement within the insulated wall may be the most important factor affecting the moisture transport into the wall. At the conclusion of this report the authors state:

"It was shown that the fibrous glass test panel accumulated moisture when air leakage through the vapor barrier occurred, and further, that this accumulation became significant only when holes were punched in the vapor barrier both in the top and bottom of the wall. It was also shown that the accumulation occurred only at the top of the panel. These facts lead to the conclusion that convection flow of moist air is of major importance in the condensation of water vapor. A separate small specimen experiment also demonstrated that the wet insulation can be dried out by applying a slight pressure in the cold chamber.

Although these experimental investigations successfully demonstrated the applicability of Takashi's air-moisture transfer theory, accurate prediction of vapor condensation and drying requires more precise knowledge of the air flow characteristics, vapor diffusion coefficient, and temperature and vapor pressure distribution across the structure. The Takashi theory should be extended for the composite wall system for its realistic application. Extended evaluation of critical air flow velocity for various parameter combinations for the outgoing air

flow system is also needed. Similar experimental studies, as performed
on the tube-type specimen are being planned at present for materials of
different air permeabilities."

Air Infiltration: Air infiltration, including wind effects, has an influence on
heating and cooling loads in a building (8). Results indicated that air infiltration
in winter can result in an average 33.4% of the heat loss from town houses while
with wind velocities of over 20 mph the heat loss can be as high as 60%.

The effect of air infiltration into a building is multiple. First, the air itself has
to be heated and secondly, in some instances, it may partially negate the benefit
of insulation in walls by introducing a cold air flow on the side of the insulation
which is normally warm. This air infiltration can be minimized by properly applied vapor/air barrier on the inside and/or a tight outer sealed sheathing.

Effective caulking of windows and doors provides the major benefit. Perforations
into and between stud spaces of walls which allow for installation of service fixtures should be well sealed. However, there are some insulation-related factors
which can help.

Filling a cavity with thermal insulation will present some resistance to air movement and will tend to reduce the effects of air infiltration. Additional benefits
can also be gained by using plastic foam sheathing materials to improve the sealing of the exterior. In this case, the overall conservation benefits measured in
actual installations utilizing plastic foam, are greater than those indicated by commonly used calculation techniques which do not correct for air infiltration. However, more experimental measurements are required to quantify fully these effects.

Air infiltration can, therefore, be seen as a significant factor which, through lack
of adequate data, demands serious and immediate attention. More information
is needed for the development of improved construction techniques. This requires improving the technology of measurement of thermal performance of systems under actual conditions.

Factors Affecting Measurement and Design

Effect of Thickness: Recent advances in thermal insulation technology, both as
a result of improvements in the measurement techniques and greater understanding of the principles of heat flow in insulating materials, have prompted con-
ceptual changes in the evaluation techniques relating to the effectiveness of insulation materials. These changes have an impact on the presentation of the
technical data for materials and systems at the greater thicknesses now in use.
The historically accepted concept of a thermal conductivity value for insulation,
which by definition is an intrinsic property obtained with materials having only
conductive modes of heat transfer governing the process, is not applicable to low
density insulations.

The change in concept is based on the fact that radiation is a significant mode
of heat transfer in low density insulations. While radiation between parallel surfaces is independent of distance, the measurement of k_{app} of materials where
the radiation mode of heat transfer is significant requires the introduction of an

additional variable, thickness (9)(10), since radiation absorption by the material is thickness dependent. This phenomenon, called the "thickness effect," is observed in materials of low density at ambient temperatures and in materials of higher density at elevated temperatures (5)(6). The effect is shown schematically in Figure 3.3. The shape and extent of the curve for a given insulation depends upon the radiation permeance of the material which is influenced by its density.

Figure 3.3: Schematic Diagram of the Effect of Thickness

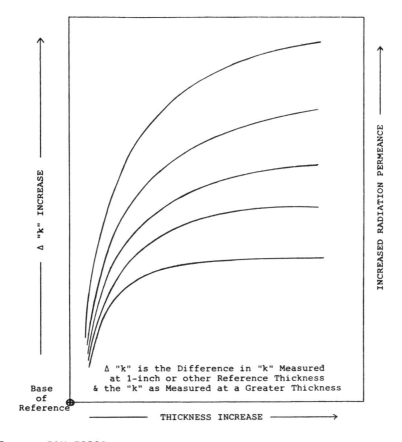

Source: BNL-50862

Low density building insulations are commonly applied at thicknesses of from 1 to 12 inches. Traditionally, measurements have been undertaken on 1 to 1.5-inch thick samples, and guarded hot plate equipment and heat flow meter equipment were developed for measurements at such thicknesses. The test methods used and the results obtained were in accordance with the technology and equipment available up to the early 1970s. However, the recent increase in knowledge of heat transport phenomena in low density thermal insulations brought about a change in the philosophy of measurements in 1973 (11).

Since that time, the test method specifications have been changed to reflect the new philosophy. These changes are now incorporated in ASTM Standard Test Methods ASTM C.177–76 and C.518–76.

Large-scale guarded hot plate equipment and heat flow meter equipment have been developed and have been used to evaluate materials at the greater thicknesses during recent years. Because the concept is relatively new, limited experience has been acquired to substantiate the accuracy of the measurements made with these larger instruments. Reference and calibration materials are necessary to verify the performance of guarded hot plate equipment or to calibrate heat flow meter equipment.

The only material available from the National Bureau of Standards to the industry as a reference material is a 1-inch thick high density (8 lb/ft^3) fiber glass. Reference materials, including low density insulations, are needed at thicknesses greater than 1 inch. The National Bureau of Standards has historically provided reference materials.

The insulation industry, in cooperation with its trade associations, independent testing laboratories, instrument manufacturers, and standards organizations, is contributing to the recognition and acceptance of full-thickness testing. The governing ASTM Standard Test Methods (STM) ASTM C 177–76 (STM for Steady-State Thermal Transmission Properties by Means of the Guarded Hot Plate Method) in its paragraph 1.6.2 and ASTM C 518–76 (STM for Steady-State Thermal Transmission Properties by Means of the Heat Flow Meter Method) in its paragraph 1.7.2 impose the requirement for "full-thickness" testing, up to a thickness for which the thermal resistivity (reciprocal of thermal conductivity) of the material does not change by more than 2% with further increases in thickness.

These Standard Test Methods are made specific to insulating materials by their incorporation by reference in the following ASTM Standard Recommended Practices and Standard Specifications:

ASTM C578-69 — Specification for Preformed, Block-Type Cellular Polystyrene Thermal Insulation.

ASTM C591-69 — Specification for Rigid Preformed Cellular Urethane Thermal Insulations (in course of revision).

ASTM C687-71 — Recommended Practice for Determination of the Thermal Resistance of Low Density Fibrous Loose-Fill-Type Building Insulation.

ASTM C726-72 — Specification for Mineral Fiber Roof Insulation Board.

ASTM C739-77 — Specification for Cellulose Fiber (Wood-Base) Loose-Fill Thermal Insulation.

ASTM C764-73 — Specification for Mineral Fiber Loose-Fill Thermal Insulation.

ASTM Cxxx — Specification for Membrane-Faced Rigid Cellular Polyurethane or Polyisocyanurate Roof Insulation Boards (in course of ballot).

ASTM Cxxx — Specification for Urea-Based and Urea-Formaldehyde Cavity Fill Insulation (in course of ballot).

ASTM C665-70, Specification for Mineral Fiber Blanket Thermal Insulation for Wood-Frame and Light Construction Buildings and ASTM C653-70, Recommended Practice for Determination of the Thermal Resistance of Low-Density Mineral Fiber Blanket-Type Building Insulation incorporate by reference both ASTM C177 and C518, but also provide for an alternative test method for thicker blanket products which allows extrapolation from measurements made on samples of partial thickness sliced from the blanket.

There is general recognition of the effect of thickness on the measurement of thermal transmission properties of building insulations. It is apparent from Figure 3.3 that the requirement for "full-thickness" testing will not impose testing at the same thickness for all insulating materials; that is, low density insulations will require testing at greater sample thicknesses than will high density insulations, to meet the criterion that the thermal resistivity not change by more than 2% with further increases in thickness.

There remains a controversy within the industry as to whether these methods, practices, and specifications for "full-thickness" testing can be implemented in the absence of availability of reference materials of lower density and substantially greater sample thickness than are presently available from NBS. The National Mineral Wood Insulation Association has provided a viewpoint as follows:

> Mineral Fiber Insulation Industry Position on
> Implementation of ASTM C.177-76 and C.518-76
>
> Major firms in the mineral fiber insulation industry have been leaders in the advances in thermal insulation technology which led to the concept of testing at full thickness as currently specified in ASTM C.177-76 and C.518-76. These studies in principles of heat flow in insulating materials have led to the rather radical change in concept. This change in concept is an outgrowth of the production of insulations which are lower in density and consequently less homogeneous. The heat flow characteristics of low density insulations are affected by sample thickness. These mineral fiber insulation industry firms supported the revised ASTM specifications as a step forward in technology.
>
> The National Mineral Wool Insulation Association, as representative of the mineral fiber insulation industry, offers an essential and complementary comment for consideration in conjunction with this report. Of great concern has been the lack of calibration standards which can be used to verify the performance of a guarded hot plate or to calibrate the more commonly available heat flow meters, the latter being a secondary dependent method. The recognized agency which should provide this calibration/referencing authority is the National Bureau of Standards. Their current equipment capability is based on the thermal conductivity concept. Their measurement capability is currently limited to a 1-inch thickness. NBS has recently taken steps to acquire the necessary equipment to provide measurement of thermal resistance of insulations at greater than 1-inch thickness. The establishment of NBS capability to make measurements at full thickness and to provide calibration and reference standards for use by other laboratories is an essential step before ASTM C.177-76 and C.518-76 can be implemented.
>
> The National Mineral Wool Insulation Association fully supports the NBS program to provide calibration capability for full thickness testing, it supports the round robin tests being implemented by ASTM, and it supports the implementation of ASTM C.177-76 and C.518-76 when NBS calibration capability is established. It supports the technical validity of the two ASTM specifications. The National Mineral Wood Insulation Association does not concur with implementation of these

specifications until NBS calibration samples are available and ASTM round robin testing has been completed.

Mean Temperature Effects: Because of equipment limitations and the practical ease of attaining mean temperatures of 75°F, thermal performance of insulations for building applications has been reported only at the above mean temperature. While selection of this temperature has practical advantages, it is not representative of temperatures encountered during winter or summer use. For example, under application conditions in winter, the outside temperature may be –10° to +10°F with the inside at 70°F, resulting in a mean temperature of 30° to 40°F.

Alternatively, for summer conditions, particularly in Southern climates, mean temperatures of 90° to 100°F can be experienced. For the lower temperatures the apparent k will be lower than that at 75°F while for the higher temperatures the opposite will be true.

Table 3.4 contains information on several typical building insulation materials to illustrate this point. Since many temperature conditions are actually encountered and a number of buildings are designed for specific uses involving temperature conditions other than 75°F, it is suggested that in the future information on thermal performance contain values covering the range of temperatures expected to occur. This could be derived from one measured temperature point combined with an accepted experimentally determined k versus temperature slope for generic material or type.

Table 3.4: Thermal Resistance Values for Different Temperature Applications (12)

Condition	Mean Temp. (°F)	Apparent Thermal Conductivity, Btu-in/hr-ft²-°F					Thermal Resistance, hr-ft²-°F/Btu-in				
		Fiber-glass*	Cellu-lose**	Molded Polystyrene***	Extruded Polystyrene***	Urea-Formal-dehyde†	Fiber-glass	Cellu-lose	Mold. Polystyrene	Extr. Polystyrene	Urea-Formal-dehyde
Winter	40	0.28	0.26	0.225	0.185	0.22	3.57	3.85	4.44	5.41	4.54
Design	75	0.316	0.27	0.25	0.200	0.24	3.16	3.70	4.0	5.00	4.17
Summer	110	0.355	0.28	0.275	0.215	0.26	2.81	3.57	3.65	4.65	3.85

*Conforming to an R11 product at 75°F.
**Loose-fill at 2.5 lb/ft³ (average results from Reference 13).
***Molded polystyrene at 1.0 lb/ft³ (averaged results). Extruded polystyrene at 2 lb/ft³.
†At 0.7 lb/ft³.

Source: BNL-50862

OTHER PROPERTIES

Fire Performance

In considering the information available to describe the fire performance of insulations it should be pointed out that the current small- and large-scale tests do not provide truly quantitative information. Results for spread of flame and intensity of smoke are given in terms of numerical ratings and are descriptive in nature. Furthermore, the various small- and large-scale tests do not necessarily provide results which correlate. Finally the major test methods now in use involve conditions and orientations of the insulation which are not those normally found for insulations as installed. It is also necessary to state that in buildings the insulation materials are normally placed between inner and outer sheathings or in systems which are highly resistant to fire. The actual information presented, therefore, should not be taken as representative of the materials in actual practice for all conditions.

The manufacturers of insulations for dwellings do not, in general, give fire data in sales brochures. They do, however, indicate compliance to pertinent Federal or other specifications which contain requirements for allowable maximum ASTM E-84 Surface Burning and Smoke Developed Values. These are pertinent values useful for comparing materials. It can be argued that as long as insulations are no less fire safe than the wood frame, they do not add further hazard to the structure when installed. However, there may be increased safety to be obtained by requiring that an insulation material comply with one or more of the following available or proposed tests.

(1) Tests for spot ignition to demonstrate that the material will not ignite or smolder under local heating conditions such as discarded cigarettes, overloaded wiring, heat sources such as recessed light fixtures, etc.

(2) Compliance to ASTM E-136 for proof of noncombustibility might permit relaxation of current requirements that such insulations do not contact sources of heat such as chimneys. This would require that wood framing also in contact with the chimney not be heated to smoldering by the increased temperature caused by insulation.

(3) Use of suggested test methods, such as the NBS Flooring Radiant Panel, or the Canadian Upside-Down E-84 Tunnel Test, to show that exposed attic insulation will not spread flame on the upper surface. These would replace current reliance on the present E-84 test which is useful for measuring surface burning on lower exposed surfaces or on vertical surfaces.

Plastic foam manufacturers have made increased use of the ICBO 8' x 12' full room fire test for evaluating the burning characteristics of plastic foams. ASTM has issued E-603-77 (a recommended practice for room fire experiments) which is intended to evaluate the burning characteristics of materials in a similar manner.

The data on insulations used for commercial and industrial applications is generally more specific. The information usually includes compliance to requirements for exterior fire exposure such as the ASTM E-108 Roofing Tests, interior fire exposure such as the UL Roof Deck Construction Tests or the Factory Mutual Calorimeter, or the concepts of limited combustibility such as the National Fire Protection Association (NFPA) No. 259 or the ASTM E-136 Combustibility Test.

At present information on toxicity of combustion gases is published only for some plastic foams. This is understandable as no accepted method of assessment is yet available. It is presumed that some data are available under a number of test conditions, but since there is no concensus as to test methods or limits, there is reluctance to provide the available data except in answer to specific requests.

It should be remembered that absolute noncombustibility of all insulation is an utopian concept not to be achieved without sacrificing many materials of good insulating quality. Moreover, a systematic approach to use of materials of low combustibility can afford adequate safety.

Corrosion

Very little information is supplied for corrosion properties. This is due in the main to there being a lack of suitable corrosion tests. However, the full impact of possible effects of the environmental conditions developing a corrosive atmosphere due to components of the insulations has not been realized until recently. In addition, in recent years many newer materials are used in building constructions and the corrosive effects of these materials have not been studied to any extent.

Moisture Absorption

In the case of moisture absorption, results are often available for one set of environmental conditions. However, some insulation materials undergo cyclic changes with environmental temperature and humidity conditions and information is required on these factors. In certain conditions, capillary action within an insulation material in a cavity could be envisioned but no test method to study this behavior is available. Some information on the effect of moisture on the thermal performance of certain plastic foams has been given. However, equivalent data for other insulating materials is not readily available.

Summary

Information, where it is available, for properties other than thermal performance is sparse, but is probably adequate for most use conditions. Certain property values for materials under actual use conditions are available for very limited or specific conditions only. The development of additional information with regard to these other properties would provide a better basis for selection of materials. More adequate test methods are required for these other properties if deemed significant for the conditions of use.

VALIDITY OF DATA

In general, it can be said that data for materials used for industrial or commercial applications is more specific and complete than that for insulations used in residential construction as shown in the latter half of this book. This appears to be the result of the more technically oriented purchaser and the more severe use requirements of insulations used in commercial or industrial applications.

The purchaser of insulations used for residential construction has been in general the builder who is satisfied in fulfilling the basic requirements established by FHA and other agencies. This rather nontechnical concern is sometimes reflected in the less than thorough attention to details of the installation of the insulation, sometimes to the point of disregarding the recommendations of the manufacturer. To ensure that design performance is achieved in applications, means should be provided at the user level to inspect and enforce the proper installation of the insulation. Inspectors will have to be made aware of factors

which govern insulation performance and proper installation procedures necessary to obtain design effectiveness. There has been, in fact, no power of enforcement in this segment of the market until recently when some building codes
have included energy conservation sections. Also, some builders have shown concern with the effectiveness of the insulation, particularly for solar heated homes.

The information reported was found generally to be consistent with accepted
limits for the generic classifications. There were only two classifications where
there were significant variations in the claims. Both of these relate to relatively
recent technologies, or at least to materials which have only recently taken an
important position in the insulation market; these are the cellulosic and some of
the cellular plastic materials. With additional experience and study, the thermal
performance of these newer materials as well as the older types of materials will
be better understood.

Mineral Fiber Blowing/Pouring Wool

The thermal resistance, R, values cited in the reported data infer apparent thermal conductivity values which can be met with the product if the density or
coverage requirements stipulated by the manufacturer are met. Much depends
on application procedures, as with all loose-fill insulation. Generally, the most
critical criterion will be the coverage, which implies a certain density and thickness requirement; if these requirements are met, the product will perform within
the limits of the generic classification of the product.

Cellulosic Blown Insulation

Since the survey provided inconclusive data from the many manufacturers, two
additional sources of information were used (13)(1). In the first study (13), a
large number of materials were measured, covering a wide range of products and
temperatures. Spray applied materials were also included. Measurements were,
in general, carried out on samples 12" x 12" having a maximum thickness of
1.5".

In the second report (1), there was a comparison of data claimed by manufacturers with measurements made at one laboratory. No details of the methods
or sample sizes were given for the manufacturer's data. However, the measurements made independently were performed on samples close to the thickness
of application.

A third report containing data is the *Building Materials Directory* published by
Underwriters Laboratories (14).

A comparison of this information is shown in Figure 3.4. This shows the wide
divergence of the results from very low claims by the manufacturers to much
higher values measured at thicknesses of use. Some of the scatter can be ex-
plained partially by the test methods used, and partially by the material itself.
The size and type of cellulosic material forming the insulation may play a part.

As discussed earlier, the extremely low values cannot be substantiated since they
approach the k of still air. The independent laboratory tests performed on the
same material showed a total variation of 55 to 63% from a "mean" value. The

ASHRAE Book of Fundamentals, Chapter 22, contains design values which propose R-value per unit thickness of the order of 3.6 hr-ft²-°F/Btu at 2.5 lb/ft³ blown density. It should be pointed out that while this value may be reasonable for material blown in attic applications, a lower R-value per unit thickness should be used for wall applications where the density is normally well above 3 lb/ft³.

Figure 3.4: Apparent Thermal Conductivity of Cellulosic Insulation

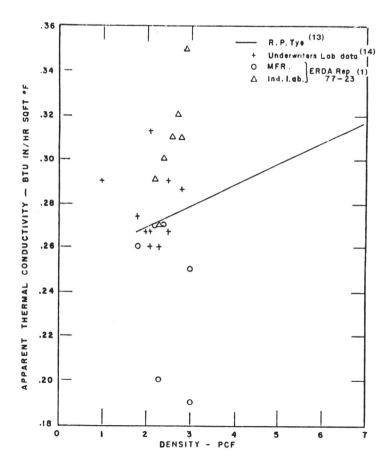

Source: BNL-50862

Mineral Fiber Batts and Blankets

The thermal performance of all mineral fiber products used in building applications is reported solely in terms of a thermal resistance, R-value, coupled with a corresponding thickness. The R-values are based upon measurements of k_{app} under the NAHB Certification Program and assume that R is directly proportional to thickness and the material fully recovers in thickness.

When mineral fiber materials are used for industrial and commercial applications they are usually higher density products, and, as stated earlier and shown in Figure 3.5, the thermal performance data are more specific.

Figure 3.5: Apparent Thermal Conductivity vs Density — Fiber Glass for Commercial Buildings

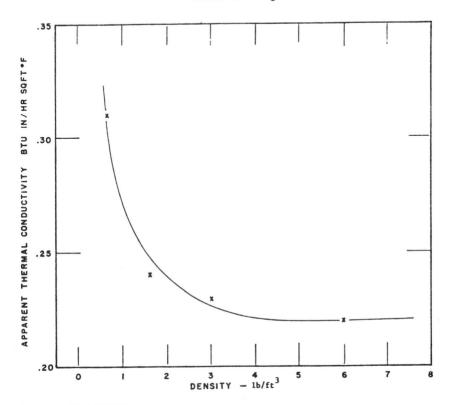

Source: BNL-50862

The k_{app} range for all products is 0.27 to 0.34 Btu-in/hr-ft^2-°F and as shown in Figure 3.2 such values can be obtained for the various products at different densities. Since fiber glass produced by different processes and rock and slag wool will have different characteristics for k_{app} versus density, density specifications for the products are inappropriate.

Urea-Based Foams

A significant range of thermal conductivities have been reported. The values of equivalent thermal resistance per unit thickness range from 5.5 to 3.5 hr-ft^2-°F/Btu. Comparisons between the values are difficult since little or no documentation specifying the test conditions was provided and in some cases the density was not given. However, for many years recommended values published for European products (15) and widely accepted there indicated that a typical R-value of 4.2

hr-ft^2-°F/Btu per inch of thickness was the same for densities in the 0.7 to 0.8 lb/ft^3 range as used in the U.S.A. This value has been substantiated by tests undertaken at NBS and NRC in Canada. Values much in excess of these numbers are clearly incorrect from the earlier basic discussions of general behavior. The European information indicates clearly the increase of thermal resistance per unit thickness with increase in density to approximately 1.8 lb/ft^3. The general behavior is thus similar to other cellular plastic materials.

A documented concern about urea-based foams is the shrinkage which takes place as the product cures. While shrinkage does not affect the thermal conductivity of the foam itself, it does detrimentally affect the thermal performance of the final installation.

Wood Fiber/Mineral Insulation Boards

The survey data on roof insulation boards both for wood fiber and mineral board products conform to the generic values of these products. These materials have been in use for many years, and meet the engineering requirements of the trade.

Polyurethane/Isocyanurate Foam Insulation Boards

For some plastic foam roof insulation boards made with urethane foam, a controversy on applicable k exists. There has been a narrowing of the range from 0.11 to 0.16 to 0.13 to 0.16 Btu-in/hr-ft^2-°F. The ASHRAE Book of Fundamentals proposes a design R-value per inch of 6 hr-ft^2-°F/Btu for urethane board products. The claims for using values at the lower end of this range are based on materials as manufactured or on claims of sealing surfaces against gas flow in or out of the foam.

The long term effects of gas movement are not fully quantified. However, it is known that the thermal performance is reduced when fluorocarbons are replaced by air. Since these materials are now used extensively in industrial/commercial applications, there will be in time some accepted long term applicable value for polyurethane foam boards protected with impervious or low permeance skins.

There also will be an applicable value for products which are not protected by impermeable skins. The urethane foam problem does not affect the data relating to composite foam mineral board panels where the reported R-values reflect the discrepancy in acceptable k of the urethane component of the panels.

Extruded Polystyrene Foam Board

Extruded polystyrene foam board has a reasonably consistent k-value of 0.20 Btu-in/hr-ft^2-°F for the commercial densities of this product. The k-values reported for extruded polystyrene foam are 5-year aged design values.

Molded Polystyrene Foam Board

Molded polystyrene foam board thermal performance does not deteriorate with age. It does, however, vary with density. The claimed k-value for 1 lb/ft^3 density foam varies from 0.23 to 0.26 Btu-in/hr-ft^2-°F. The claimed k-value for 2 lb/ft³ density ranges from 0.20 to 0.23. Additional quantification of actual performance is needed.

MATERIAL PROPERTIES

Material properties for the various insulation materials are presented and discussed in this section. Tables 3.5 through 3.12 present values of the material properties for each generic classification of materials, based upon the assessment of available data.

Fiber Glass

Fiber glass insulation is made of thin glass fibers felted in batts or nodules. It falls into two major classifications as a result of different manufacturing processes:

> Process A — The fibers are produced by melting glass marbles into primary fibers which are attenuated into relatively long fibers.
>
> Process B — This process, known as the rotary process, produces fibers by flowing molten glass into a spinning perforated disc, thus projecting glass fibers. This process produces shorter fibers. For both processes, the fibers are sprayed with a binder and collected into wool-like mats that have a high air content, forming a low-density material (typically less than 1 lb/ft³) with good thermal resistance. Fiber glass is cut into batts or blankets of various thicknesses and lengths which have a derived R-value of about 3.2 per inch thickness. An important factor in thermal performance of a batt or blanket product is that it should recover to its design thickness.

Glass fiber loose-fill insulations (for blowing and pouring) are usually produced by hammer milling glass fiber blanket material, thereby retaining the bonded fiber quality which insures good loft and thus a high R per pound ratio. A typical R-value for this form is 2.2 per inch thickness at 75°F.

Both loose-fill and batt or blanket forms of fiber glass insulation are permeable to water vapor to the extent of over 100 perm-inch. Water absorption is typically no more than 1% by weight, by ASTM C553-70, and no capillarity is apparent in these materials.

Fiber glass itself is an inorganic, noncombustible material, but flammable, organic binders are used in the production of batts and blowing wool. ASTM E-84 yields an approximate rating for the material with binder of flame spread: 15 to 20; fuel contributed: 5 to 15; smoke developed: 0 to 20.

Facings on fiber glass building insulation usually consist of an asphalt-coated-kraft or foil-kraft paper laminate which is a flammable surface. The facing must not be exposed to open flames or temperatures exceeding 180°F. Any burning of facings or organic binders used could produce thermal degradation products which are toxic.

Table 3.5: Fiber Glass

Material Property	Value	Test Method
Density	0.6-1.0 lb/ft^3	—
Thermal conductivity (k factor)	varies with density	—
Thermal resistance (R value)*	3.16 hr-ft^2-°F/Btu (batt) 2.2 hr-ft^2-°F/Btu (loose-fill)	ASTM C-518, C-653
Water vapor permeability	>100 perm-in	—
Water absorption	<1% by weight	ASTM C-553-70
Capillarity	none	—
Fire resistance	non-combustible	ASTM E-136
Flame spread	15-20	ASTM E-84
Fuel contributed	5-15	ASTM E-84
Smoke developed	0-20	ASTM E-84
Toxicity	toxic fumes could develop due to binder combustion	—
Effect of age		
Dimensional stability	none (batt) settling (loose-fill)	—
Thermal performance	none	—
Fire resistance	none	—
Degradation due to:		
Temperature	none below 180°F	—
Cycling	none	—
Animal	none	—
Moisture	none	—
Fungal/bacterial	does not promote growth	—
Weathering	none	—
Corrosiveness	noncorrosive	Federal HH-I-558D
Odor	none	ASTM C-553

*Per 1 inch of thickness at 75°F; thickness derived from R19 and R11 products for 6-inch and 3.5-inch thickness respectively.

Source: BNL-50862

Fiber glass batt insulation does not appear to settle or shrink with age, but loose-fill may settle. Other properties of the material, such as thermal performance and resistance to fire, are reportedly unaffected by age and temperature cycling at normal installed temperatures. Fiber glass does not promote bacterial or fungal growth, and provides no sustenance to vermin. Insulation products made from fiber glass are noncorrosive (Federal Spec. HH-I-558D) and have no objectionable odor (ASTM C553-Sec. 16).

Rock and Slag Wool

The overwhelming majority of this product manufactured in the U.S. is made from steel, copper, or lead slag as opposed to natural rock as raw material which is used extensively in Europe. The slag is melted using coke as a fuel, then spun into fibers by pouring the molten material onto rotating discs. The fibers are attenuated with steam, and rapidly cooled to room temperature.

In rock or slag wool insulation the fibers do not have the resiliency of glass. After packaging, the product may fail to recover to the design thickness, thus resulting in lower than design R-values. The fibers are sprayed with a phenolic resin, which serves as a binder, compressed, and cured by passing through an oven. The resulting "slabs" are cut to desired sizes to make batts. Another additive is mineral oil which serves to seal the surface against dust production and give water-repellency.

Table 3.6: Rock and Slag Wool

Material Property	Value	Test Method
Density	1.5–2.5 lb/ft^3	—
Thermal conductivity (k factor) at 75°F	0.31–0.27 Btu-in/ft^2-hr-°F (batts) 0.34 Btu-in/ft^2-hr-°F (loose-fill)	ASTM C-177
Thermal resistance (R value) per 1" thickness, at 75°F	3.2–3.7 hr-ft^2-°F/Btu (batts) 2.9 hr-ft^2-°F/Btu (loose-fill)	ASTM C-177
Water vapor permeability	>100 perm-in	—
Water absorption	2% by weight	—
Capillarity	none	—
Fire resistance	noncombustible	ASTM E-136
Flame spread	15	ASTM E-84
Fuel contributed	0	ASTM E-84
Smoke developed	0	ASTM E-84
Toxicity	none	—
Effect of age		
Dimensional stability	none (batt), settling (loose-fill)	—
Thermal performance	none	—
Fire resistance	none	—
Degradation due to:		
Temperature	none	—
Cycling	none	—
Animal	none	—
Moisture	transient	—
Fungal/bacterial	does not support growth	—
Weathering	none	—
Corrosiveness	none	—
Odor	none	—

Source: BNL-50862

Rock wool batts and blowing wool are produced with densities in the range of 1.5 to 2.5 lb/ft^3, and reported unit thermal resistances (R-values) of 3.2 to 3.7 hr-ft^2-°F/Btu-in at 75°F (k-factor 0.31 to 0.27 Btu-in/hr-ft^2-°F) for batts, and 2.9 at 75°F (k-factor 0.34) for blowing wool. Water vapor permeability is reported (survey) to be >100 perm-in, and water absorption up to 2% by weight. Rock wool exhibits little or no capillary action.

Made from rock or slag, and melting above 1200°C, the base material of batts or blowing wool is noncombustible, but binders added to the wool may be flammable. Flame spread is reported (survey) to be less than 25 (by ASTM-84). Asphalt-coated or foil-laminated kraft paper may be used as a vapor retardant facing on batts, and should be considered flammable. Rock wool may be used as a high temperature insulation with a temperature limit of about 1000°C. Because the insulation is noncombustible, no toxic gases are generated. Burning of facings or organic binders could produce toxic thermal degradation products.

Properties such as dimensional stability, thermal performance, and fire resistance are reportedly unaffected by age, temperature cycling, or weathering. Thermal conductivity is affected by moisture content, but the change is transient and the material returns to its original properties upon drying. Rock wool does not support the growth of fungus, bacteria, or vermin, exudes no odor and is non-corrosive.

The thermal properties of the material are affected by "shot" content, pieces of slag that spun off as particles rather than fibers. Higher apparent thermal conductivity with density increase is due to high shot content, and this content is usually reported by individual products.

Cellulose

As described earlier, cellulose insulation is manufactured from waste paper, primarily used newsprint or virgin wood fiber, by shredding and milling to produce a fluffy, low density material. Chemicals are then added to provide resistance to fire, water absorption, and fungal growth.

Table 3.7: Cellulose

Material Property	Value	Test Method
Density	2.2–3.0 lb/ft^3	—
Thermal conductivity		ASTM C-177
(k factor) at 75°F	0.27–0.31 Btu-in/ft^2-hr-°F	and C-518
Thermal resistance (R value)		
per 1" of thickness at 75°F	3.7–3.2 hr-ft^2-°F/Btu	—
Water vapor permeability	high	
Water absorption	5–20% by weight	ASTM C-739
Capillarity	not known	—
Fire resistance	combustible	ASTM E-136
Flame spread	15–40	ASTM E-84
Fuel contributed	0–40	ASTM E-84
Smoke developed	0–45	ASTM E-84
Toxicity	develops CO when burned	—
Effect of age		
Dimension stability	settles 0–20%	—
Thermal performance	not known	—
Fire resistance	inconsistent information	—
Degradation due to:		
Temperature	none	—
Cycling	not known	—
Animal	not known	—
Moisture	not severe	—
Fungal/bacterial	may support growth	—
Weathering	not known	—
Corrosiveness	may corrode steel, aluminum, copper	ASTM C-739
Odor	none	ASTM C-739

Source: BNL-50862

Typical applied cellulose insulation density is in the range of 2.2 to 3.0 lb/ft^3 when pneumatically blown or poured in place in attics and somewhat higher in wall cavities. Accepted thermal resistance values are in the range of 3.7 to 3.2 ft^2-hr-°F/Btu per inch for this density range. Compaction due to vibration and settling under its own weight results in a decrease in R-value in two ways: loss in thickness, and an increase in k_{app} due to the increase in density.

According to ASTM C739-73, weight gain from water absorption should not exceed 15%. Properly treated cellulosic materials do not exceed this specification, but poor quality control and improper selection of flame retardant chemical may result in greater water absorption. Some information on permeability of the material to water vapor has been reported (1).

Large amounts (up to 25% by weight) of chemicals, primarily boric acid, aluminum sulfate, ammonium sulfate, and calcium sulfate, are added to provide flame retardancy. Separation of these additives from the insulation material has been reported (1) although there are no standard specifications (ASTM, Federal, or NCIMA) or requirements for nonseparation of fire retardant additives.

Values for flame spread, fuel contributed, and smoke developed according to ASTM E-84 varied from product to product but generally fall within the specifications for Class 1 materials. These chemicals may cause corrosion on metals such as steel, aluminum, and copper (1).

Cellulose insulation is a material in which fungal and bacterial growth may be a problem. Bacteria and fungus that degrade cellulose as a thermal insulation material are present everywhere. Chemicals are usually added which inhibit the development of these problems. Cellulose has no odor, and its properties are unaffected by cycling, with the possible exception of moisture cycling.

Expanded Polystyrene Foam

Polystyrene foam insulation is manufactured in two forms: extruded, and molded expanded bead, as discussed in the previous chapter.

Foam produced by the extrusion process has a more consistent density, uniform appearance, and greater compressive and tensile strength than that produced by the molding process. Extruded density is usually in the range of 2.0 lb/ft^3. The reported (from the manufacturer survey) k-factor is 0.12 Btu-in/hr-ft^2-°F as manufactured, but as the air diffuses in, the k-factor rises to 0.20 Btu-in/hr-ft^2-°F. This value with an equivalent R per 1-inch thickness of 5 hr-ft^2-°F/Btu is normally accepted for this material in use.

Extruded polystyrene shows a permeability to water vapor of 0.6 perm-in when tested by ASTM-C355-64 and a volumetric water absorbance of 0.7% (21.8% by weight) by ASTM-C2842-69. There is no apparent capillary action by polystyrene.

Molded polystyrene is made to have densities in the range of 0.8 to 1.8 lb/ft^3. Variations of about 10% from the average density can be found in a piece of molded polystyrene due to the molding process. Thermal conductivity of this material is directly proportional to density, and is usually in the range of 0.23 to 0.26 Btu-in/hr-ft^2-°F. This value does not change with age. The R-value for molded polystyrene is lower than the R-value for extruded polystyrene since the former has air in the cells while the latter has a mixture of air and fluorocarbon. Water vapor permeability for the molded material is reported to be 1.2 to 3.0 perm-in by ASTM-C355, and water absorption less than 2% by volume by ASTM-C272.

Table 3.8: Expanded Polystyrene Foam

Material Property	Value	Test Method
Density	0.8–2.0 lb/ft³	–
Thermal conductivity (k factor)	0.20 Btu-in/ft²-hr-°F (extruded)	ASTM C-177
	0.23–0.26 Btu-in/ft²-hr-°F (molded)	and C-518
Thermal resistance (R value) per 1″ thickness at 75°F	5 hr-ft²-°F/Btu (extruded)	–
	3.85–4.35 hr-ft²-°F/Btu (molded)	
Water vapor permeability	0.6 perm-in (extruded)	ASTM D-2842-
	1.2–3.0 perm-in (molded)	69 and C-355
Water absorption	<0.7% by vol (extruded)	
	<0.02% by vol (extruded)	ASTM D-2842-
	<4% by vol (molded)	69 and C-272
	<2% by vol (molded)	
Capillarity	none	–
Fire resistance	combustible	ASTM E-136
Flame spread	5–25	ASTM E-84
Fuel contributed	5–80	ASTM E-84
Smoke developed	10–400	ASTM E-84
Toxicity	develops CO when burned	–
Effect of age		
Dimensional stability	none	–
Thermal performance	k increases to 0.20 after 5 yr (ext.)	–
	none (mold.)	
Fire resistance	none	–
Degradation due to:		
Temperature	above 165°F	–
Cycling	none	–
Animal	none	–
Moisture	none	–
Fungal/bacterial	does not support	–
Weathering	direct exposure to UV light degrades polystyrene	–
Corrosiveness	none	–
Odor	none	–

Source: BNL-50862

The advantage of polystyrene boardstock used as frame sheathing in building construction is in providing insulation over the whole building frame, thus minimizing the effect of the more conductive structural members. An even more important benefit of this use is the reduction of air infiltration into a building; this foam sheathing provides a better air seal than conventional sheathing. Plastic foam sheathing materials are nonstructural with low nail holding ability.

Other properties of polystyrene insulation are independent of the manufacturing process. Polystyrene is combustible, and in use must be covered with a flame resistant covering such as gypsum board. It must also be protected from direct exposure to ultraviolet light, which causes dusting and yellowing. Insulating properties, however, are not affected by short-term exposure to ultraviolet light. The maximum service temperature of polystyrene is 165°F; exposure to higher temperatures will cause the plastic to soften. There is no effect of cycling or weathering on the insulation in the service temperature range. Polystyrene does not promote the growth of fungus or bacteria, and contains nothing of food value for animals. This insulation has no odor, and is noncorrosive.

Polyurethane and Polyisocyanurate Foams

Polyurethane and polyisocyanurate foams are fluorocarbon-blown materials available in both precast board stock and spray-in-place forms. These foams have a thermal conductivity (k-factor) of 0.11 to 0.15 Btu-in/ft^2-hr-°F at 75°F when new, and a density of 2.0 lb/ft^3. Closed cell content of these rigid foams is approximately 90%; cells are filled with fluorocarbon blowing agent. The fluorocarbon vapor in the cells has a significantly lower thermal conductivity than air, which explains the low k-factor of the material.

There is some question about the validity of published "aged" values of thermal conductivity for urethane or isocyanurate foam. These range from 0.13 to 0.16 Btu-in/hr-ft^2-°F. It is known that "as manufactured" foam will have values of 0.11 to 0.12 Btu-in/hr-ft^2-°F, but that the thermal conductivity will increase as the foam ages and air replaces the fluorocarbon gas in the cells.

This process is reduced or eliminated when a relatively air tight facing is used on the foam. The ASTM Standard Specification for Rigid Preformed Cellular Urethane Thermal Insulation, C-591-69, shows values of 0.16 to 0.17 for material aged over 300 days with an initial value of 0.11 to 0.12 Btu-in/hr-ft^2-°F.

The advantage of polyurethane or polyisocyanurate board stock used as frame sheathing in building construction is in providing insulation over the whole building frame, thus minimizing the effect of the more conductive structural members. A major manufacturer of polyisocyanurate foam sheathing specifies vent strips to allow escape of water vapor which has penetrated the inner face vapor barrier. This may lessen the benefits of reduced air infiltration.

Because of the high closed cell content, water absorption and permeability are very low; permeability is typically 2 to 3 perm-in.

Polyurethane and polyisocyanurate foams are flammable and must be covered with a fire retardant material when used for thermal insulation in most applications. Certain polyisocyanurate foams have been approved for exposed use in certain industrial/commercial buildings. Typical burning characteristics for polyurethane are a flame spread of 30 to 50, fuel contributed value of 10 to 25, and smoke developed of 155 to over 500. For polyisocyanurate, the flame spread is less than 25, fuel contributed is less than 5, and smoke developed is 55 to 200.

Polyurethane and polyisocyanurate foams show dimensional change upon curing and aging. The degree of expansion or shrinkage is related to conditions of temperature and humidity and the duration of exposure to extreme conditions. For polyurethane, results of ASTM-D-2126 Procedure F (160°F and 100% RH) indicate a change in volume of up to 12% after 14 days. For polyisocyanurate, results with this same test indicate a 3% change in volume after 14 days. Most compositions of these foams begin to decompose above 250°F. Polyurethane and polyisocyanurate foams are resistant to fungal and bacterial growth.

Table 3.9: Polyurethane and Polyisocyanurate Foams

Material Property	Value	Test Method
Density	2.0 lb/ft^3	—
Closed-cell content	90%	ASTM C-591-69
Thermal conductivity	0.16–0.17 Btu-in/ft^2-hr-°F*	ASTM C-177
(k factor)	0.13–0.14 Btu-in/ft^2-hr-°F**	and C-518
Thermal resistance (R value)	6.2–5.8 hr-ft^2-°F/Btu*	
per 1" thickness at 75°F	7.7–7.1 hr-ft^2-°F/Btu**	—
Water vapor permeability	2–3 perm-in	—
Water absorption	negligible	—
Capillarity	none	—
Fire resistance	combustible	ASTM E-136
Flame spread	30–50 polyurethane	
	25 polyisocyanurate	ASTM E-84
Fuel contributed	10–25 polyurethane	
	5 polyisocyanurate	ASTM E-84
Smoke developed	155–500 polyurethane	
	55–200 polyisocyanurate	ASTM E-84
Toxicity	produces CO when burned	—
Effect of age		
Dimensional stability	0–12% change	ASTM D-2126
Thermal performance	0.11 new, 0.17 aged 300 days	—
Fire resistance	none	—
Degradation due to:		
Temperature	above 250°F	—
Cycling	not known	—
Animal	none	—
Moisture	limited information available	—
Fungal/bacterial	does not promote growth	—
Weathering	none	—
Corrosiveness	none	—
Odor	none	—

*Aged and unfaced or spray applied.
**Impermeable skin faced.

Source: BNL-50862

Urea-Based Foams

This insulation is used primarily for retrofit applications and is normally manufactured on site and applied directly. An aqueous solution of a urea-formaldehyde-based resin is combined with an aqueous solution foaming agent which includes an acid catalyst or hardening agent and compressed air.

Recently the National Bureau of Standards has studied the insulating properties performance of this material (2) and the information presented herein consists largely of the major results suggested in their report.

Table 3.10: Urea-Formaldehyde and Urea-Based Foams

Material Property	Value	Test Method
Density	~2.5 lb/ft^3 wet	*
	~0.6–0.9 lb/ft^3 dry	*
Closed-cell content	0.7–80%	—
Thermal conductivity		
(k factor)	0.24 Btu-in/hr-ft^2-°F	ASTM C-177-76
Thermal resistance (R value)		
per 1″ thickness at 75°F	4.2 hr-ft^2-°F/Btu	—
Water vapor permeability	4.5–100 perm-in	
	(50% RH, 73°F)	ASTM C-355
Water absorption	32% by wt (0.35% vol), 95% RH	
	18% by wt (0.27% vol), 60% RH,	
	68°F	—
	180–3,800% by wt (2–42% vol),	
	immersion	
Capillarity	slight	—
Fire resistance	combustible	ASTM E-136
Flame spread	0–25	ASTM E-84
Fuel contributed	0–30	ASTM E-84
Smoke developed	0–10	ASTM E-84
Toxicity	no more toxic than burning	
	wood fumes	—
Effect of age		
Dimensional stability	1–4% shrinkage**	
	4.6–10% shrinkage***	ASTM D-2126C
	30–45% shrinkage†	ASTM D-2626-66
Thermal performance	no change	—
Fire resistance	—	—
Degradation due to:		
Temperature	decomposes at 415°F	—
Cycling	no damage after 25 freeze-thaws	—
Animal	not a feed for vermin	—
Moisture	not established	—
Fungal/bacterial	does not support growth	ASTM G-21-70
Weathering	—	—
Corrosiveness	—	—
Odor	may exude formaldehyde	
	until cured	—

*Weigh a foam-filled plastic bag of known volume.
**In 28 days due to curing.
***At 100°F, 100% RH for 1 week.
†At 158°F, 90–100% RH for 10 days.

Source: BNL-50862

It is concluded that the physical properties which are agreed upon in the available literature include density, mechanical strength and water absorption. However, a significant range of thermal conductivities has been reported. From basic considerations of general behavior, European experiences, and measurements both at NBS and NRC in Canada, an R-value of 4.2 hr-ft^2-°F/Btu is recommended as a reliable value at the density of use, 0.7 to 0.8 lb/ft^3.

For the remaining properties of interest, the NBS reports that there are sufficient data available from which the performance of the material may be adequately evaluated, and, further, that some of the data in the literature are contradictory.

In particular, concern was expressed with regard to the present understanding of dimensional stability of urea-formaldehyde-based foams and its resistance to high temperature and high humidity. The NBS conducted preliminary tests which indicated that the material performance with respect to shrinkage and resistance to high temperature and humidity may be suspect. Until time-based studies reveal the in-service performance of this material, the question of durability will remain unanswered.

It should be pointed out that European practice, now also being applied in Canada, is to derate the overall calculated or measured thermal performance of systems containing urea-based foams up to 40% in order to allow for shrinkage associated factors. The recent HUD specifications recommend derating to 72% of claimed R-value for design purposes (16). Further study is required on the various parameters which can affect the thermal performance of systems containing this insulation.

Perlite

Expanded perlite can be produced with densities in the range of 2 to 11 lb/ft³. This material has a pellet-like form which contains countless tiny glass-sealed air cells that account for its thermal insulation properties. Nonflammable silicone treatment increases its resistance to water penetration and perlite is claimed to be water repellent and impervious to moisture. Being inorganic, perlite is rot, vermin, and termite resistant, and is noncombustible. It softens at temperatures between 890° and 1100°C, and melts between 1280° and 1350°C. The thermal conductivity of loose-fill perlite varies with density. This density dependence is illustrated in Figure 3.6.

Figure 3.6: Thermal Conductivity of Perlite (17)

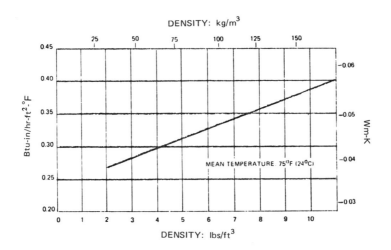

Source: BNL-50862

Expanded perlite is also mixed with Portland cement to form a lightweight insulating concrete. Density is varied by controlling the perlite/cement ratio, and a range of 20 to 40 lb/ft^3 is typical. Perlite concrete, which may be precast in a number of shapes or cast-in-place, possesses sufficient mechanical strength to be load-bearing at high densities. It has a k-factor of 0.51 to 2.00 Btu-in/hr-ft^2-$°$F, k increasing with increasing density.

Table 3.11: Perlite

Material Property	. . . . Value		Test Method
	Loose-Fill	Perlite Concrete	
Density	2–11 lb/ft^3	20–40 lb/ft^3	—
K_{app} at 75$°$F	0.27–0.40 Btu-in/ft^2-hr-$°$F	0.50–0.93 Btu-in/ft^2-hr-$°$F	ASTM C-177
Thermal resistance (R value)			
per 1" thickness at 75$°$F	3.7–2.5 hr-ft^2-$°$F/Btu	2.0–1.08 hr-ft^2-$°$F/Btu	—
Water vapor permeability	high	high	—
Water absorption	low	—	—
Capillarity	—	—	—
Fire resistance	noncombustible	noncombustible	ASTM E-136
Flame spread	0	0	ASTM E-84
Fuel contributed	0	0	ASTM E-84
Smoke developed	0	0	ASTM E-84
Toxicity	none	none	—
Effect of age			
Dimensional stability	none	none	—
Thermal performance	none	none	—
Fire resistance	none	none	—
Degradation due to:			
Temperature	none under 1200$°$F	none under 500$°$F	—
Cycling	none	none	—
Animal	none	none	—
Moisture	none	none	—
Fungal/bacterial	does not promote growth	does not promote growth	—
Weathering	none	none	—
Corrosiveness	none	none	—
Odor	none	none	—

Source: BNL-50862

Vermiculite

Vermiculite is exfoliated to many times its original value when heated rapidly to high temperatures (700$°$ to 1000$°$C) by steam which forms and forces the laminae apart. By controlling the degree of exfoliation, a density range of, typically, 4 to 10 lb/ft^3 can be produced in the expanded material. The lower density material has an average particle size of 6.5 mm and is commonly used as loose-fill insulation and concrete aggregate.

Higher density material is used as plaster aggregates and for high temperature applications, as particle sizes are smaller (1.5 to 4.5 mm), air voids between particles are smaller and convection is minimized. The thermal conductivity of exfoliated vermiculite is typically 0.33 to 0.41 Btu-in/hr-ft^2-$°$F at ambient temperatures, which translates to R-values of 3.0 to 2.4 per inch.

Table 3.12: Vermiculite

Material Property	. . . . Value . . . . Loose-Fill	Vermiculite Concrete	Test Method
Density	4–10 lb/ft^3	20–60 lb/ft^3	–
K$_{app}$ at 75°F	0.33–0.41 Btu-in/ft^2-hr-$^\circ$F	0.59–0.96 Btu-in/ft^2-hr-$^\circ$F	ASTM C-177
Thermal resistance (R value)			
per 1" thickness at 75°F	3.0–2.4 hr-ft^2-$^\circ$F/Btu	1.7–1.0 hr-ft^2-$^\circ$F/Btu	–
Water vapor permeability	high	high	–
Water absorption	none	none	–
Capillarity	none	none	–
Fire resistance	noncombustible	noncombustible	ASTM E-136
Flame spread	0	0	ASTM E-84
Fuel contributed	0	0	ASTM E-84
Smoke developed	0	0	ASTM E-84
Toxicity	none	none	–
Effect of age			
Dimensional stability	none	none	–
Thermal performance	none	none	–
Fire resistance	none	none	–
Degradation due to:			
Temperature	none below 1000°F	none below 1000°F	–
Cycling	none	none	–
Animal	none	none	–
Moisture	none	none	–
Fungal/bacterial	does not promote growth	does not promote growth	–
Weathering	none	none	–
Corrosiveness	none	none	–
Odor	none	none	–

Source: BNL-50862

Vermiculite is also mixed with Portland cement and sometimes sand to produce vermiculite concrete. Densities of this material usually range from 20 to 60 lb/ft^3, with higher densities resulting in higher thermal conductivities which range from 0.59 to 0.96 Btu-in/hr-ft^2-$^\circ$F (R of 1.7 to 1.0 per inch).

Vermiculite is treated to ensure water repellency. It is noncombustible and melts at 1315°C. Being an inorganic material, it is resistant to rot, vermin, and termites, and is not affected by age, temperature, or humidity. Vermiculite is chemically inert, therefore, noncorrosive and exudes no odors.

Insulating Concrete

The density of insulating concrete is proportional to the degree of foaming. The range is 12 to 88 lb/ft^3. As is the case with perlite and vermiculite concretes, thermal resistance decreases sharply and compressive strength increases, with increasing density. For example, a material which is strong enough for compressive loading purposes may have a density of 40 lb/ft^3 and a thermal conductivity of 1.17 Btu-in/hr-ft^2-$^\circ$F (R = 0.85 per inch) while a material with density of 25 lb/ft^3 cannot be used for load-bearing purposes but has a thermal conductivity of about 0.83 Btu-in/hr-ft^2-$^\circ$F (R = 1.2 per inch). The thermal conductivity of insulating concrete increases with increasing moisture content (18)(19).

Table 3.13: Insulating Concrete

Material Property	Value	Test Method
Density	12–88 lb/ft^3	—
Thermal conductivity	1.17 Btu-in/ft^2-hr-$^\circ$F (40 lb/ft^3)	
(k factor) at 75°F	0.83 Btu-in/ft^2-hr-$^\circ$F (25 lb/ft^3)	ASTM C-177
Thermal resistance (R value)	0.85 hr-ft^2-$^\circ$F/Btu (40 lb/ft^3)	
per 1″ thickness at 75°F	1.2 hr-ft^2-$^\circ$F/Btu (25 lb/ft^3)	—
Water vapor permeability	varies with density	—
Water absorption	—	—
Capillarity	none	—
Fire resistance	noncombustible	ASTM E-136
Flame spread	0	ASTM E-84
Fuel contributed	0	ASTM E-84
Smoke developed	0	ASTM E-84
Toxicity	none	—
Effect of age		
Dimensional stability	none	—
Thermal performance	none	—
Fire resistance	none	—
Degradation due to:		
Temperature	none below 1000°C	—
Cycling	none	—
Animal	none	—
Moisture	none	—
Fungal/bacterial	does not support growth	—
Weathering	below 30 lb/ft^3 must be pro- tected from frost	—
Corrosiveness	none	—
Odor	none	—

Source: BNL-50862

Insulating concrete is highly resistant to moisture penetration; however, materials with a density of less than 30 lb/ft^3 may be damaged by frost and must be protected when used in exterior applications. The material is noncombustible and can be exposed to temperatures as high as 1000°C without damage. Because of its inorganic nature, it does not support the growth of fungus or vermin. Shrinkage is less than 0.05% between the fully water-saturated and completely dry stages. Precast autoclave-cured material is very stable and has no tendency to deteriorate with age.

Reflective Surfaces

Reflective materials act as insulations by reflecting incident radiant heat energy, rather than by reducing conduction as conventional bulk insulations do. In a vertical cavity, a wall, for example, about 60% of heat transfer is due to radiation, while in loft and under floor spaces radiation contributes about 50 and 70% to heat transfer respectively.

Aluminum foil, the most common reflective insulation material, is effective in reducing such losses by 90% when applied to one side or both of a cavity. However, aluminum foil has relatively little effect on conduction or convection heat losses. Thus, aluminum foil would be of greatest benefit in applications were convective and conductive heat losses are relatively small (such as under floor spaces).

The reflective insulations consist of a varying number of reflective air-filled cells. The thermal resistance of the system is governed by the number of cells and the direction of heat flow. As a system, it performs very effectively in reducing the radiative heat transfer. It does require, however, a reflective air space, a condition which is not always adequately met. The foil itself is noncombustible and an excellent vapor barrier.

Aluminum foil insulation has a low mass and heat capacity, so that buildings employing it will heat rapidly, and it takes up minimal space in a wall. Foil is not affected by age or temperature and will not support the growth of fungus, bacteria, or vermin, but the surfaces may be affected by water vapor. If dulling of the reflective surface by oxidation or dusting takes place, the effectiveness will be reduced.

Other Materials

In concluding this section on materials, reference is made to other materials which are only in the experimental or developmental stage for consideration as thermal insulations in buildings. These include:

(1) Foamed gypsum whereby the base product is combined with glass fiber and a fluid-blowing agent. This material has the advantage of very good fire resistance properties and the promise of good thermal performance at low densities.

(2) Foamed asbestos, an European product, has similar attributes to foamed gypsum. However, it does have possible health hazards.

(3) Polyvinyl-chloride foams are closed cell cellular plastics having a similar range of densities as the present polystyrene and polyurethanes. In general, their performance characteristics will be somewhat similar to present materials.

CERTIFICATION AND INSPECTION PROGRAMS

Mineral fiber batt and blanket insulation products may carry a third party certification label under the National Association of Home Builders Research Foundation Certification Program. Under this program, inspectors from NAHB-RF select samples from each producing plant measure dimensions after the insulation is removed from the compressed package, and test for thermal performance in their own research laboratory as described above. Labels attesting to NAHB-RF certification are contributed by NAHB-RF and are permitted only on individual items produced at specific plants where NAHB-RF tests show continuing compliance.

Manufacturers of other building insulation materials, including mineral fiber loose-fill, are now considering similar certification and inspection programs for their respective products.

REFERENCES

(1) Anderson, R.W., and Wilkes, P., ERDA 77-23 UC-95d (January, 1977).

(2) Rossiter, W.J., Jr., Mathey, R.G., Burch, D.M., and Pierce, E.T., NBS Technical Note 946 (July, 1977).

(3) Ball, G.W., Hurd, R., and Walker, M.G., *J. Cell. Plast.,* 9, 66 (1970).

(4) "ECON-1 How to Determine Economic Thickness of Thermal Insulation," Thermal Insulation Manufacturers Association, Mt. Kisco, NY (1973).

(5) *Thermal Conductivity,* Vol. I, Chap. 6, Ed. Tye, R.P., Academic Press, London (1969).

(6) Pelanne, C.M., *J. Thermal Insul.,* 1, 48 (1977).

(7) Kusuda, T. and Ellis, W., NBSIR Technical Report (April, 1975).

(8) Harrje, D.T., "Retrofitting: Plan, Action and Early Results Using Townhouses at Toms River, NJ," Report #29, Princeton University (1976).

(9) Larkin, B.K. and Churchill, S.W., *JAIChE,* 5 (4) 467 (1959).

(10) Jones, T.T., *Proc. VIIth Thermal Conductivity Conference,* NBS Sp. Publ. 302, 737 (1967).

(11) *Heat Transmission Measurements in Thermal Insulations,* ASTM STP 544, ASTM Stand., Philadelphia (1974).

(12) American Society of Heating, Refrigerating, and Air-Conditioning Engineers, Inc., *ASHRAE Handbook of Fundamentals* (1977).

(13) Tye, R.P., *J. Test Eval.,* 2 (3), 176 (1974).

(14) *Building Materials Directory,* Underwriters Laboratories, Inc. (1977).

(15) *The Chartered Guide,* Section A3, Institute of Heating and Ventilating Engineers, Cadegan Square, London.

(16) Housing and Urban Development, "Usage of Materials," 74.

(17) Perlite Institute, Inc., "Perlite Technical Data Sheet/No. 2-4," (1977).

(18) Valore, R.C., Jr., *J. Amer. Concrete Inst.,* 28 (5), 502 (1956).

(19) Tye, R.P., and Spinney, S.C., "Thermal Conductivity of Concrete: Measurement Problems and Effect of Moisture," IIF-IIR, Commission B1, Washington (1976).

RESIDENTIAL BUILDING
INSULATION ASSEMBLIES

The material in this chapter was excerpted from a report prepared by Brookhaven National Laboratory with the assistance of Dynatech R/D Company (BNL-50862).

BUILDING INSULATION

In use, building insulation must be installed in or on the building envelope and the heat loss or gain through the envelope becomes much more complex than for insulation alone. The structural members are usually more conductive than the envelope itself and the system must limit or account for that greater heat flow. The structural members are typically wood, steel, or masonry and the envelope insulation may be located between the framing members or on the interior or exterior of those members. Placement of insulation outside the frame reduces the effect of short circuits of heat flow through the structural members.

Table 4.1 presents a simplified classification of insulation as used in elements of the building envelope. Residential and industrial/commercial categories are separated because of major differences in building construction. (Industrial/commercial building insulation is described in the next chapter.) Wood frame construction is only a small portion of the industrial/commercial segment, whereas it constitutes the major form of residential construction, Table 4.2.

Many of the applications listed in Table 4.1 are not widely used. However, it is important to describe some of these lesser used applications because they offer solutions to some of the more difficult to insulate types of construction, and they are often applicable to retrofit as well as new construction. Also, the applications could become more widely used as increasing energy costs make them more economically attractive.

Table 4.3 shows the relative importance of various building categories as reported in December, 1976.

Table 4.1: Insulation Assemblies

Composite Assembly	Insulation Used	
	Residential	Industrial/Commercial
Roof/Ceiling		
Roof truss/joists	Mineral fiber batt or loose fill, cellulose loose fill	Fiber batt, loose fill
Overdeck	Foam board, wood fiber board	Foam board, mineral or wood fiber insulation board, insulating concrete
Over-roof membrane	Foam board	Foam board
Underdeck	Mineral fiber batt, wood fiber board	Mineral fiber batt, spray-in-place foam and wood fiber insulation board
Cathedral ceiling	Mineral fiber batt, foam or wood fiber board	—
Walls		
Cavity wall	Mineral fiber batt or loose fill, foam, cellulose loose fill	Mineral fiber batt, foam, cellulose
Insulated frame-sheathing	Foam board, wood fiber board	Foam board, fiber batt
Interior wall substrates	Foam board	Foam board, fiber batt
Stucco base	—	Foam board
Panels	—	Panels with foam, wood fiber, or mineral board, mineral fiber batts encased in metal or mineral faces
Floor		
Masonry slab substrate or perimeter insulation	Foam board	Foam board
Wood framed floor	Mineral fiber batt	Mineral fiber batt
Foundation/Wall		
Exterior application	Foam board	Foam board
Interior wall substrate	Foam board, mineral fiber batt	Foam board, mineral fiber batt

Note: Mineral loose fill insulations such as perlite and vermiculite or foam pellets are used to fill voids in concrete blocks.

Source: BNL-50862

Table 4.2: 1973 Typical Builder Practices in Single Family Homes (1)
(Percent of Housing Units)

| | . . . . Region . . . . | | | |
	North-east	North Central	South	West	US Total
Roof structures					
Roof trusses	59.2	69.5	72.1	72.3	70.0
Rafters	37.6	28.1	24.7	20.2	25.8
Other (and no answer)	3.2	2.4	3.2	7.4	4.1
Total	100.0	100.0	100.0	99.9	99.9
Exterior wall structures					
2″ x 4″ studs	95.1	97.1	80.3	90.1	89.8
Solid masonry	0.5	0.1	12.9	7.0	6.4
Other (and no answer)	4.3	2.8	6.8	2.9	3.8
Total	99.9	100.0	100.0	100.0	100.0
Floor framing					
Wood joist/beam	89.9	88.6	45.6	59.9	61.2
Other (and no answer)	10.0	2.4	4.5	8.2	5.6
Not applicable (concrete)	0.0	9.0	49.9	31.9	33.3
Total	99.9	100.0	100.0	100.0	100.1
Foundation construction					
Concrete block	28.8	45.0	33.8	2.4	28.7
Poured concrete	67.8	51.5	60.7	95.6	67.3
Other (and no answer)	3.4	3.5	5.5	2.0	4.0
Total	100.0	100.0	100.0	100.0	100.0
Foundation type					
Full/partial basement	51.8	71.4	20.4	24.7	36.8
Crawl space	3.0	7.0	16.5	25.4	15.0
Slab-on-grade	43.3	18.0	60.0	42.4	44.3
Other (and no answer)	1.9	3.5	3.1	7.4	4.0
Total	100.0	99.9	100.0	99.9	100.1

Source: BNL-50862

Table 4.3: Construction Potentials (2) (December 1976)

	% Total ft^2
Residential buildings	
Single family	52.1
Multi family	12.3
Total housekeeping residential	64.4
Total nonhousekeeping residential	1.2
Total residential buildings	65.6
Industrial/Commercial buildings	
Total commercial buildings	16.0
Total manufacturing buildings	5.4
Total educational and science buildings	4.3
Total hospital and other health treatment buildings	2.6
Total amusement, social and recreational buildings	1.7

(continued)

Table 4.3: (continued)

	% Total ft²
Total public buildings	1.4
Total religious buildings	1.1
Total miscellaneous nonresidential buildings	1.9
Total industrial/commercial buildings	34.4
Total buildings	100.0

Source: BNL-50862

The importance of the residential building area is apparent. Almost two-thirds of major new construction, additions, and alterations were in residential buildings. In fact, single family units alone accounted for over half of major construction and remodeling square footage in the United States. One and two family units are usually similar in construction. These two categories accounted for 53.5% of the total. The remaining 46.5% of major new construction, additions, and alteration projects utilize a wide variety of methods of construction, yet there are many similarities in basic construction methods. The basic methods will be described in the chapter on Industrial/Commercial building insulation.

Residential construction is standardized to a great extent and, therefore, provides a consistent basis for analysis and evaluation. Table 4.3 shows the primary construction practices for the major structural portions of single family homes which will be insulated. The following discussion presents the areas that must be considered.

HEAT LOSSES IN RESIDENTIAL BUILDINGS

In considering energy conservation within buildings, residential structures are a major area primarily because of the number of structures involved (approximately 65% of the total buildings). This is especially true when the subject of retrofitting of the estimated 25 million under-insulated dwellings (defined as less than the HUD Minimum Property Standards of R-11 walls and R-19 ceilings) which were deemed likely candidates for reinsulation (3).

The principal heat losses in residential buildings are through the ceiling, the walls, the floors, the windows and doors, and through infiltration. Table 4.4 shows the level of some of these losses in an uninsulated house and a well-insulated house. The problems of heat gain during the summer air conditioning periods, as shown in Table 4.5, are similar. The relationships remain about the same.

Ceilings

Attic or ceiling insulation results in a significant reduction in heat loss. This very significant improvement is due to the fact that an uninsulated ceiling has practically no resistance to heat flow (less than an R of 3 when the two air film resistances and the plaster are considered). In an uninsulated house, the heat loss through the ceiling amounts to more than 40% of the total heat loss. This problem is the first to be corrected, and is often simple to correct.

Table 4.4: Heat Loss Distribution for Heating a Typical Residence (4)
(thousand Btu)

	. . . . Uninsulated.		. . . .Well-Insulated	
	Peak Hour	Day	Peak Hour	Day
Ceiling	18.4 (42%)	338.8 (43%)	2.8 (12%)	53.9 (14%)
Wall	7.0 (16%)	177 (22%)	2.8 (12%)	68.6 (17%)
Glass				
Conduction and convection	9.3 (22%)	130 (16%)	9.3 (40%)	130.0 (33%)
Infiltration	8.4 (20%)	144.7 (19%)	8.4 (36%)	144.7 (36%)
Total load	43.1	790.5	23.3	397.2

Note: Peak hour is the hour of the day which corresponds to the highest thermal load.

Source: BNL-50862

Table 4.5: Heat Loss Distribution for Cooling a Typical Residence (4)
(thousand Btu)

	. . . . Uninsulated.		. . . .Well-Insulated	
	Peak Hour	Day	Peak Hour	Day
Ceiling	5.5 (18%)	84.0 (22%)	1.6 (7%)	22.9 (8%)
Wall	2.9 (10%)	44.0 (12%)	1.1 (5%)	14.9 (6%)
Glass				
Conduction and convection	2.1 (7%)	29.2 (8%)	2.1 (9%)	29.2 (10%)
Radiation	14.9 (49%)	117.6 (31%)	14.9 (60%)	117.6 (41%)
Infiltration	5.0 (16%)	101.3 (27%)	5.0 (19%)	101.2 (35%)
Total load	30.4	376.1	24.7	285.8

Note: Peak hour is the hour of the day which corresponds to the highest thermal load.

Source: BNL-50862

Walls

A wall consists of a number of resistive surfaces and materials which produce a moderate thermal resistance (approximately R-5). Thus, the addition of insulation into uninsulated walls does not produce the dramatic improvement obtained with the addition of insulation in uninsulated ceilings. The insulation of the walls of older houses is more difficult; it can be achieved by blowing or otherwise injecting insulation into the cavities. Significant savings can be achieved by insulating basement walls.

Floors/Foundations

The insulation of floors, which are not listed in Tables 4.4 and 4.5, is subject to much debate. Definitely, floors over ventilated unheated crawl spaces should be insulated. Past practice has been to heat the crawl space partially with the heat losses from the heating air ducts and other basement heat sources. Frequently, floors over unheated basements have been uninsulated, since it was felt that the rising heat from the furnace would minimize the heat loss from the living area.

In Table 4.6 taken from the Arkansas Study (5), it can be seen that insulating the floors can result in significant savings of energy.

Table 4.6: The Arkansas Story*—Heat Loss Comparison Chart** (5)

	Energy Conservation Construction	FHA Minimum Property Standards
	. . . . (Btu)	
Floor	3,179	8,722
Walls	4,411	6,757
Ceiling	2,041	4,320
Windows and doors	3,050	13,131
Infiltration	3,007	7,548
Subtotal	15,688	40,478
Duct loss	471 (3%)	6,072 (15%)
Total Btu heat loss	16,159	46,550

Note: Total heat loss reduction = 65%.

*Joint project by Arkansas Power & Light Company and others to demonstrate method for Energy Conservation Report by Owens-Corning Figerglas.
**From Arkansas Power & Light Load Calculation Forms at 70° Temperature Difference Heating and 25° Temperature Difference Cooling.

Source: BNL-50862

Perimeter insulation of foundation walls is an alternative to floor insulation primarily for unvented crawl spaces. For slab-on-grade construction, the perimeter edge of the slab should be insulated in most geographical areas. The furnace should be supplied with outside air by ducts to reduce air demand from the inside, thus decreasing air infiltration.

Windows and Doors

While strictly not insulation materials, the overall effect of windows and doors is included in this analysis. They present an entirely different problem. Because they break the continuity of the walls, they produce many peripheral heat loss areas, some of which are not corrected by doubling the glass panes. In general, heat loss calculations do not consider the solar collecting capabilities of the windows. Selecting location and protection of the windows should result in a significant improvement in efficiency. Protective draperies can be used effectively in controlling the heat losses or gains from the windows. The use of storm windows has the advantage of reducing air infiltration in addition to the double pane benefit.

Table 4.6 also illustrates the total benefits in conserved energy which could be obtained by modifying existing standard construction techniques to take full advantage of the use of thermal insulation. Ceilings had 12" of fiber glass insulation, wall cavity insulation was increased from 3½" to 5½" in thickness, floors were insulated with 6" of insulation and 1½" of rigid urethane foam perimeter insulation was installed around concrete slabs. Window and door area was reduced from 166 to 106 ft², while caulking was used extensively to reduce infiltration. The comparison of heat losses from the house with these above conditions with that of a comparable one insulated to FHA Minimum Property Standards shows that overall savings due to added insulation are truly significant.

While this example applies to new construction, it does illustrate problem areas of existing buildings and the significance of their solution by the correct use of insulation in the various sectors.

Heat Loss Calculations

Consideration should be given to the effect of added amounts of insulation. Using recommended ASHRAE methods, the overall coefficient of heat transmission (U) was calculated for typical 2 x 4 stud (16" on center) walls with various amounts of insulation. These calculations use the recommended 20% framing correction and a 75°F mean design temperature. The results are shown in Table 4.7.

Adding an R-7 insulation to the cavity of an uninsulated wall produces a 56% reduction in conductive heat loss. The addition of R-13 insulation to the cavity of an uninsulated wall produces a 66% reduction in the conductive heat loss. This is 24% better than the use of R-7 cavity insulation.

Similarly, when both plastic foam sheathing and R-13 insulation are added to an uninsulated wall, there is a 74% reduction in calculated conductive heat loss, which is a 24% improvement over a wall with R-13 cavity insulation. This illustrates that, percentage wise, initial installation of insulation provides the most significant reduction in conductive heat flow.

Additional amounts of insulation yield progressively less reductions in heat loss. This is true of all insulations and all insulated areas (roof/ceilings, walls, floors/foundations). However, the value of added insulation should be judged by the quantity and dollar value of energy saved and the pay-back on the insulation investment. Additional amounts of insulation can be economically justifiable when energy costs and the severity of the climate are taken into account via a proper economic analysis.

The following Thermal Performance Guidelines recommended by the National Association of Home Builders (NAHB) are examples of approaches which consider the economic justification of added insulation to establish a base for its recommendations:

> "Retrofitting Existing Housing for Energy Conservation: An Economic Analysis" by the National Bureau of Standards and the FEA,
> "Optimum Insulation Thickness in Wood-Framed Homes" by the U.S. Department of Agriculture, and
> "In the Bank. . .or Up the Chimney" by the Department of Housing and Urban Development.

Table 4.7: The Effect of Increasing Insulation in an Opaque Wall

	No Cavity Insulation		R-7 Cavity Insulation		R-13 Cavity Insulation		R-13 Cavity Insulation plus Plastic Foam Sheathing	
	Through Framing	Through Cavity	Through Framing	Through Cavity	Through Framing	Through Cavity	Through Framing	Through Cavity
. . . . . . Thermal Resistance $(hr\text{-}ft^2\text{-}{}^\circ F/Btu$ at $75\,{}^\circ F$ mean temperature)								
Outside air film	0.17	0.17	0.17	0.17	0.17	0.17	0.17	0.17
Siding (average)	0.40	0.40	0.40	0.40	0.40	0.40	0.40	0.40
½" Wood fiber sheathing	1.32	1.32	1.32	1.32	1.32	1.32	–	–
1" Extruded polystyrene sheathing	–	–	–	–	–	–	5.0	5.0
2" x 4" Stud	4.35	–	4.35	–	4.35	–	4.35	–
Air space	–	0.94	–	0.81	–	–	–	–
Mineral fiber insulation	–	–	–	7.00	–	13.00	–	13.00
½" Gypsum board	0.45	0.45	0.45	0.45	0.45	0.45	0.45	0.45
Inside air film	0.68	0.68	0.68	0.68	0.68	0.68	0.68	0.68
Total resistance at section (R)	7.37	3.96	7.37	10.83	7.37	16.02	11.05	19.7
. . . . Heat Transmission $(Btu/hr\text{-}ft^2\text{-}{}^\circ F)$								
U at section	0.1357	0.2525	0.1357	0.0923	0.1357	0.0624	0.0905	0.0508
Framing correction	x 20%	x 80%	x 20%	x 80%	x 20%	x 80%	x 20%	x 80%
	0.0271	0.2020	0.0271	0.0738	0.0271	0.0499	0.0181	0.0406
U of wall	0.2291		0.1009		0.0770		0.0587	
U improvement (over no cavity insulation)	–		0.1282		0.1521		0.1704	
Percent improvement (incremental)	–		56		24		24	
Percent improvement (total)	–		56		66		74	

Source: BNL-50862

RESIDENTIAL ROOF/CEILING ASSEMBLIES

Table 4.8 describes in more detail the predominant roof/ceiling assemblies for residential construction. The predominant roof structures are the roof truss and the rafter type of construction.

Table 4.8: Roof/Ceiling Assemblies (1)

	. . . . Region				
	North-east	North Central	South	West	US Total
Roofing material					
Shingles/shakes	99.3	95.8	93.0	88.0	93.2
Built-up	0.0	0.5	0.8	4.7	1.4
Other and no answer	0.7	3.7	6.3	7.2	5.4
Total	100.0	100.0	100.1	99.9	100.0

(continued)

Table 4.8: (continued)

| | . . . . Region. . . . . | | | | |
	North-east	North Central	South	West	US Total
Roof structures					
Roof trusses	59.2	69.5	72.1	72.3	70.0
Rafters	37.6	28.1	24.7	20.2	25.8
Other and no answer	3.2	2.4	3.2	7.4	4.1
Total	100.0	100.0	100.0	99.9	99.9
Ceiling surface					
Gypsum board (½" or more)	80.9	88.9	82.2	82.4	83.4
Other and no answer	19.1	11.1	17.9	17.7	16.5
Total	100.0	100.0	100.1	100.1	99.9

Source: BNL-50862

Roof Truss/Rafter Cavity Applications

The rafter has traditionally been 2" x 6" or 2" x 8" joists spaced 16" on center. The materials used for roof/ceiling applications of this construction are the traditional inexpensive loose fill and batt mineral fiber, cellulose fiber, and vermiculite insulations, since there are seldom any thickness restrictions. These systems are used in new construction and retrofit applications.

As shown in Figure 4.1, there are three methods of installing blanket insulation in ceilings. Loose fill fiber insulation can be installed as shown in Figure 4.2. There are, however, several less predominant roof/ceiling assemblies. These include cathedral ceilings, rafters spaced 24" on center, overdeck applications, and insulated roof membranes.

Figure 4.1: Installation of Blanket Insulation in Ceilings (6)

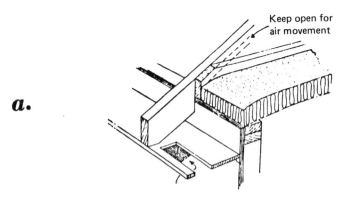

a.

Keep open for air movement

(continued)

Figure 4.1: (continued)

b.

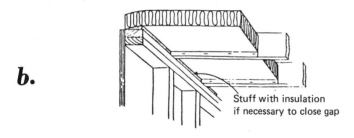

Stuff with insulation
if necessary to close gap

c.

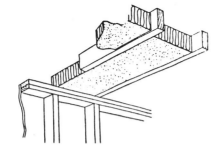

(a)   Laying faced insulation in from above when the ceiling
      finish material is in place
(b)   Stapling from below
(c)   Installing unfaced pressure-fit blankets with a separate
      vapor barrier wedged between ceiling joists

Source: BNL-50862

Figure 4.2: Typical Loose Fill Rafter Application

ATTIC AREA

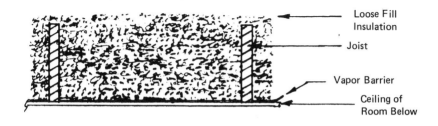

Loose Fill
Insulation

Joist

Vapor Barrier

Ceiling of
Room Below

Source: BNL-50862

Cathedral Ceiling Substrate Application

Rigid insulation boards are useful in cathedral ceilings because of limited space in which to position mineral fiber or loose fill insulation. One effective method of improving the insulation of the cathedral ceiling is to use cellular plastic foams. This application has essentially three variations. It can be used to cover an exposed deck while leaving the beams exposed for aesthetics as shown in Figure 4.3a. This is applicable to both new construction and to retrofit.

A second variation, an application for new construction where there are no exposed beams, is shown in Figure 4.3b. Increased thermal performance, which cannot be met by the traditional materials in the rafter cavity, can be attained by application of an appropriate thickness of plastic foam. For instance, if R-25 insulation is desired in a cathedral ceiling constructed of 2" x 6" rafters and filled with R-19 mineral wool batts, 1" of plastic foam can be used to bring the total R value (insulation alone) to R-24. The insulating values of the inside and outside air films, the ½" plywood deck and the interior ½" gypsum board will bring the total R value to R-25, even when considering the thermal shorts created by the rafter.

Figure 4.3: Cathedral Ceiling Construction

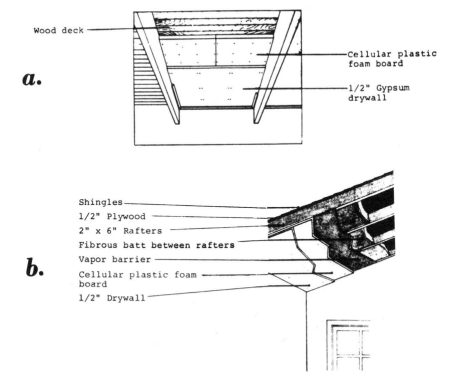

Source: BNL-50862

A third variation is a retrofit application and is very much like the second variation. Here the cellular plastic foam is applied directly to the existing cathedral ceiling and a layer of ½" drywall is then applied and finished.

Overdeck Applications

When an owner or builder wishes to have a cathedral ceiling with exposed deck and beam for aesthetic reasons, he can utilize an overdeck application such as that shown in Figure 4.4. The desired thermal performance dictates the required thickness of any given type of rigid cellular plastic foam boards. This application is particularly suitable to new construction, but is also viable as a retrofit application if the need and decision to reshingle has been established. The Asphalt Roofing Manufacturers Association recommends that plywood be installed over the foam board before roofing application.

Figure 4.4: Cathedral Ceiling Overdeck Construction

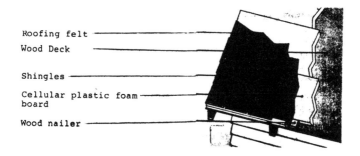

Roofing felt
Wood Deck
Shingles
Cellular plastic foam board
Wood nailer

Source: BNL-50862

Insulated Roof Membrane

Insulated roof membrane assemblies are applicable to built-up roofs. Table 4.8 shows that the usage of built-up roofs in residential construction is very small, only 1.4% in 1973. This system is used more extensively for Industrial/Commercial buildings.

24-Inch on Center Rafter

The use of wood rafters 24-inches on center is desirable from the viewpoint of reducing overall heat loss through the roof. This spacing permits more efficient use of insulation by reducing the framing which short-circuits heat flow.

RESIDENTIAL WALL ASSEMBLIES

Table 4.9 describes the predominant wall components used in residential construction. The primary structure consists of 2" x 4" stud framing on 16" centers with a gypsum board interior finish, fiberboard sheathing, and a variety of exterior finish materials.

Table 4.9: Wall Components (1)

	North-east	North Central	South	West	US Total
	. . . . .Region				
Interior wall surface					
Gypsum board (½" or more)	80.9	88.9	82.9	82.4	83.4
Other and no answer	19.1	11.1	17.2	17.7	16.5
Total	100.0	100.0	100.1	100.1	99.9
Exterior wall structures					
2" x 4" studs	95.1	97.1	80.3	90.1	89.8
Solid masonry	0.5	0.1	12.9	7.0	6.4
Other and no answer	4.3	2.8	6.8	2.9	3.8
Total	99.9	100.0	100.0	100.0	100.0
Exterior wall sheathing					
½" fiberboard (all grades)	49.9	81.8	57.4	31.9	56.3
None	6.4	4.4	2.1	35.8	10.5
½" and ⅜" plywood	33.9	6.7	1.3	16.2	9.7
½" gypsum board	4.5	3.6	15.6	3.6	9.1
Other and no answer	5.3	3.4	23.5	12.5	14.3
Total	100.0	99.9	99.9	100.0	99.9
Principal exterior finish materials					
Brick veneer	10.3	20.8	51.1	13.1	31.4
Aluminum	38.0	34.7	10.3	0.0	16.6
Wood siding	2.9	5.7	10.3	9.5	8.4
Stucco	0.0	0.0	0.7	33.3	7.9
Hardboard	0.9	15.0	1.3	12.5	6.8
Plywood	8.8	2.5	1.7	17.1	6.4
Concrete	0.0	0.0	6.1	2.3	3.2
Other and no answer	39.0	21.5	18.4	12.3	19.4
Total	99.9	100.2	99.9	100.1	100.1

Source: BNL-50862

Insulated Frame-Cavity Assemblies

The most popular and economic stud cavity insulations for use with 2" x 4" framing on 16" centers are 3½" R-11 and 3½" R-13 mineral fiber batts. When used in a conventional wall and allowing for a 20% framing as estimated by ASHRAE, the R-11 batts provide a U range of 0.076 to 0.084 Btu/hr ft^2 °F and the R-13 batts provide a range of 0.070 to 0.077. The range is dictated by the exterior finish. This translates to overall calculated opaque wall resistance of about R-12 to R-14. In order to improve the opaque wall thermal resistance, the conventional wall must be changed. There are essentially three alternatives, one of which is accomplished by changing to 2" x 6" stud wall constructions and two of which can be accomplished with cellular plastic foam with the conventional 2" x 4" stud wall construction.

As described in the chapter on for retrofit applications, other insulation materials are used: blown-in mineral fiber, cellulose fiber and urea-formaldehyde and urea-based foams. These systems lend themselves to wall retrofit applications where there is no insulation in the cavity. The R-value for the insulation material between the frame is approximately R-11 for blown-in mineral fiber, R-12 for cellulose, R-15 for urea-based foams (3½"), and R-19 for polyurethane/polyisocyanurate, but it should be noted that this last-mentioned insulation is rarely used for this application in residential construction. The conductive thermal performance for urea-formaldehyde foam-in-place insulation is presented in Table 4.10. Similar calculations are presented in the next section for R-11 and R-13 insulations. The use of foam-in-place urea-based foam requires careful application. There is an appreciable amount of water released during curing. Shrinkage and humidity-aging are two important factors which affect long-term performance (7)(8). Urethane spray-on-foams have high thermal resistance, but are expensive and require complicated application.

Table 4.10: Insulated Frame-Cavity Assemblies (Conductive Thermal Performance)*

| ½" Sheathing | . . . . .Siding . . . . | | | | | |
	½" Hard Board	⅜" Plywood	½" Aluminum	½" Beveled 8" Lapped Wood	4" Brick Veneer + ¾"–1" Air Space	½" Stucco
Regular density	R 14.4	R 14.14	R 14.32	R 14.57	R 15.43	R 13.8
wood fiberboard	U 0.0694	U 0.0707	U 0.0698	U 0.0686	U 0.0648	U 0.0724
Intermediate density	R 14.29	R 14.05	R 14.21	R 14.46	R 15.3	R 13.65
wood fiberboard	U 0.07	U 0.0712	U 0.0703	U 0.0691	U 0.0653	U 0.0738
Exterior grade	R 13.55	R 13.3	R 13.49	R 13.73	R 14.6	R 12.9
plywood	U 0.0738	U 0.0752	U 0.0741	U 0.0728	U 0.0684	U 0.0775

*These values do not take into account the effects of any shrinkage of urea-formaldehyde foam on thermal performance. See References 7 and 8, which indicate that a 6% foam shrinkage results in a 28% decrease in R-value for the wall assembly. R- and U-values calculated for 75°F mean temperature.

Source: BNL-50862

The toxicities of most of these foams examined here have been tested by the manufacturers. During the curing stage, the spray-on urea-formaldehyde foam does release some unpleasant odor. When foam is installed properly, the odor disappears within a few days. The products of combustion were also examined for polyurethane foam and urea-formaldehyde foam. Studies by manufacturers indicate that the toxicity of products of combustion is not worse than those released by burning the same weight of red oak.

Insulated Frame-Foam Sheathing Assemblies

An insulated frame-sheathing assembly, Figure 4.5, consists of an interior finish, a stud cavity with or without insulation in the cavity, exterior insulating sheathing, and exterior siding. This type of assembly was not widely used for residential applications, as is indicated in Table 4.9 for 1973 data. However, as the cost of energy increases, the use of insulated sheathing assemblies has become more economically attractive and, therefore, more widely accepted. This has already been observed over the last few years by a growth from a negligible amount in 1975 to about 10% of the sheathing market in 1977.

Figure 4.5: Insulated Frame-Sheathing

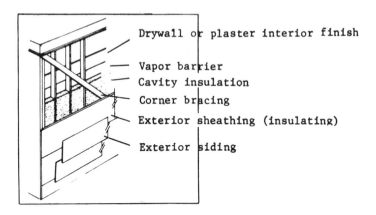

Source: BNL-50862

The overall conductive thermal performance of the opaque wall using an insulated sheathing assembly is calculated utilizing recommended ASHRAE methods. Representative R-values (at 75°F) for the various wall components are given in Table 4.11 and calculated thermal resistances are presented in Tables 4.12 and 4.13 for the indicated combinations of wall components. For these calculations, the interior finish was assumed to be ½" gypsum drywall board and wood studs were assumed to be 2" x 4"—the most popular dimension. Also, a 20% framing factor, as estimated by ASHRAE for studs on 16" centers, was used.

The cavity insulation was assumed to be R-11 and R-13 for the calculations in Tables 4.12 and 4.13, respectively. A total of 96 combinations were examined and the results for conventionally sheathed walls indicate a calculated opaque wall thermal resistance of R-10.8 to R-13.1 with an R-11 cavity insulation and R-11.9 to R-14.3 with an R-13 cavity insulation. Use of insulated sheathing improves the opaque wall resistance to a range of R-14.3 to R-17.4 with R-11 cavity insulation and R-15.5 to R-18.3 for R-13 cavity insulation. The ranges are due to the type of exterior finish and the type of sheathing material used.

The component R-values used for these cases were for 75°F mean temperature. This allows for comparison of various assemblies. ASHRAE Handbook of Fundamentals gives data on the effect of mean temperature on R-values of insulations. Most of the data are given for industrial insulations, but these values could be used to estimate effect of mean temperature on R-values of building insulations if data for building insulations are not available. The R-value of an insulation increases as temperature decreases. In winter conditions where the mean temperature is usually less than 75°F mean, the insulation performs better than predicted from an R at 75°F mean. Conversely, in warm weather or summer conditions, the performance of the insulation is less than predicted from 75°F mean data.

Table 4.11: R-Value of Wall Components

| Sheathing | . . . . . . R-Value* . . . . . . | |
	Through Framing	Between Framing
Inside air	0.68	0.68
½" drywall	0.45	0.45
2" x 4" wood studs	4.35	0
Mineral fiber with vapor barrier		
R-11 batts	0	11.0
R-13	0	13.0
Foam-in-place		
Urea-formaldehyde (3½")	0	15.2
Loose fill		
Cellulose fiber insulation, 3.5 pcf	0	12.0
Spray-on sheathing		
Polyurethane (3")	0	18.78
Regular density fiberboard (½")	1.32	1.32
Intermediate density fiberboard (½")	1.22	1.22
Exterior grade plywood (½")	0.62	0.62
Extruded, expanded polystyrene foam (1")	5.0	5.0
Molded, expanded polystyrene foam		
No. 1 density (1")	3.85	3.85
Molded expanded polystyrene foam		
No. 1.5 density (1")	4.17	4.17
Polyurethane-isocyanurate foam with		
impermeable facings (¾")	5.4	5.4
Polyurethane-isocyanurate foam with		
permeable facing or without facing (¾")	4.7	4.7
Siding		
Hardboard (½")	0.67	0.67
Plywood (³⁄₈")	0.47	0.47
Aluminum (½")	0.61	0.61
½" Beveled, 8" lapped wood sidings	0.81	0.81
4" Brick veneer plus ¾" or 1" air space	0.44 + 1.1	0.44 + 1.1
Stucco facing (½")	0.15	0.15
Outside air	0.17	0.17

*Expressed as $hr\text{-}ft^2\text{-}°F/Btu$ at 75°F mean temperature.

Source: BNL-50862

The availability of data over a temperature range of 25°F to 105°F would make it possible for a designer to utilize data which more closely correspond to the anticipated conditions of use. Accurate estimates can be made using approximate mean temperatures, but for extremely accurate calculations an iterative process is necessary. Such design calculations can give a more accurate projection of heating and cooling loads of a building and aid in sizing of heating and cooling equipment. In such calculations, other factors such as air infiltration should be considered, as they could be significant.

The calculated performance as presented in Tables 4.11 and 4.12 considers only the heat loss through conduction. Some sheathing materials have special tongue and groove edge treatment so that the wall composite is most likely tighter than the conventional wall. Air infiltration reduction has been observed in both occupied and unoccupied homes using these sheathings. Further, additions of insulation at the sheathing location may improve the thermal performance of the system by reducing the ΔT across the cavity which in turn reduces any remaining convection currents in the cavity.

Table 4.12: Insulated Frame-Sheathing Assemblies (Conductive Thermal Performance at 75°F Mean Temperature)

R-11 Mineral Fiber Batts with Vapor Barrier

Sheathing		½" Hardboard	⅜" Plywood	½" Aluminum	Siding Beveled ½" / 8" Lapped Wood	4" Brick Veneer + ¾"-1" Air Space	½" Stucco
½" Regular density wood fiberboard	R	12.18	11.95	12.11	12.33	13.1	11.6
	U	0.0821	0.0837	0.0826	0.0811	0.0763	0.0862
½" Intermediate density wood fiberboard	R	12.06	11.84	12.0	12.22	13.0	11.5
	U	0.0826	0.0884	0.083	0.0819	0.0769	0.087
½" Exterior grade plywood	R	11.4	11.18	11.34	11.56	12.36	10.82
	U	0.0877	0.0894	0.0882	0.0865	0.0809	0.0924
1" Extruded polystyrene foam	R	16.1	15.9	16.0	16.2	17.0	15.5
	U	0.0622	0.0630	0.0624	0.0616	0.0589	0.0644
1" Expanded polystyrene foam, 1 pcf	R	14.9	14.7	14.8	15.0	15.8	14.3
	U	0.0672	0.0682	0.0675	0.0666	0.0633	0.0698
1" Expanded polystyrene foam, 1.5 pcf	R	15.21	15.0	15.15	15.36	16.12	14.66
	U	0.0657	0.0667	0.0660	0.0651	0.0620	0.0682
¾" Urethane-isocyanurate impermeable facing	R	16.5	16.3	16.43	16.65	17.39	15.96
	U	0.0606	0.0614	0.0608	0.0600	0.0575	0.0627
¾" Urethane-isocyanurate permeable facing	R	15.77	15.55	15.7	15.91	16.67	15.22
	U	0.0634	0.0643	0.0637	0.0628	0.0599	0.0657

Source: BNL-50862

Table 4.13: Insulated Frame-Sheathing Assemblies (Conductive Thermal Performance at 75°F Mean Temperature) R-13 Mineral Fiber Batts with Vapor Barrier

Sheathing	½" Hardboard		³/₈ Plywood		½" Aluminum		½" Beveled 8" Lapped Wood		4" Brick Veneer + ¾"–1" Air Space		½" Stucco	
	R	U	R	U	R	U	R	U	R	U	R	U
½" Regular density wood fiberboard	13.28	0.0753	13.05	0.0766	13.2	0.0757	13.44	0.0744	14.26	0.0701	12.69	0.0788
½" Intermediate density wood fiberboard	13.17	0.0759	12.94	0.0773	13.1	0.0763	13.33	0.075	14.15	0.0706	12.57	0.0796
½" Exterior grade plywood	12.13	0.0824	12.25	0.0816	12.40	0.0805	12.64	0.0791	13.48	0.0742	11.87	0.0842
1" Extruded polystyrene foam	17.3	0.0577	17.1	0.0585	17.3	0.0579	17.5	0.0572	18.3	0.0548	16.8	0.0596
1" Expanded polystyrene foam, 1 pcf	16.1	0.0622	15.9	0.0630	16.0	0.0624	16.2	0.0616	17.0	0.0587	15.5	0.0644
1" Expanded polystyrene foam, 1.5 pcf	16.43	0.0608	16.31	0.0613	16.39	0.0610	16.58	0.0603	17.0	0.0587	15.86	0.0630
¾" Urethane-isocyanurate impermeable facing	17.75	0.0563	17.62	0.0567	17.69	0.0565	17.81	0.0560	18.67	0.0536	17.18	0.0582
¾" Urethane-isocyanurate permeable facing	17.0	0.0588	16.79	0.0595	16.94	0.059	17.15	0.0583	17.93	0.0558	16.56	0.0603

Source: BNL-50862

Not all plastic foam sheathings provide benefits of reduced air infiltration. Some manufacturers recommend use of vent strips, or have vents molded into the board to reduce possible condensation problems. Application procedures may be slightly different for the various wall assemblies. For instance, the plastic foam sheathing requires 1" x 4" let-in bracing or metal cross or angle bracing at the corners (Figure 4.5).

The manufacturer of impermeable faced isocyanurate foam sheathing specifies vent strips on the top or bottom plates (in colder climates), and 6-mil polyethylene film vapor barrier. The manufacturer of extruded, expanded polystyrene foam sheathing and the manufacturers of conventional sheathings specify warm-side vapor barriers in accordance with the National Mineral Wool Insulation Association recommendations, e.g., polyethylene film or asphalt-laminated paper facing. For some molded, expanded polystyrene foam sheathing systems, no vapor barrier is specified on the warm side. Because plastic foams are not nail base materials, the nails used for the sidings need to be long enough to penetrate studs $\frac{3}{4}$", and adjustments may be necessary in the application techniques for the exterior siding.

The installation methods are fairly straightforward. The glass fiber insulation comes with either asphalt-laminated paper/foil facing or nonfaced, friction-fit batts. The former can be stapled on either the side or the face of the studs, and the latter needs a polyethylene film to act as a vapor barrier. The plastic foam sheathings are stapled or nailed to the studs with roofing nails, then the sidings are nailed to the studs.

The fire protection rating of wall composites with various plastic foam sheathings are similar. The critical element in the wall component is the $\frac{1}{2}$" gypsum drywall which is commonly used in residential construction. Most plastic foam sheathing manufacturers specify $\frac{1}{2}$" gypsum drywall or equivalent as the recommended fire barrier, e.g., a 15-minute finish rating by ASTM E-119 test. It is also required by the three model codes. The products of combustion of the plastic foam types described here were tested extensively.

Some advantages and disadvantages of plastic foam sheathing boards are presented below.

Advantages

1. Simple application—The application is not unlike that of conventional sheathing products.
2. Lightweight and easy to handle—Foam board sheathing products are lighter in weight than conventional sheathing products. Lifting, placing, and nailing in place is easily accomplished.
3. Reduction of air infiltration—Most applications of rigid plastic foam sheathing products result in a reduction of air infiltration. All foam products with joints on the studs will reduce air infiltration. Those products which have an edge joining system such as tongue and groove or ship lap need not be joined on studs to reduce air infiltration. Those products recommending vent strips do not provide the full benefit of air infiltration reduction.

4. High insulation value—The thermal resistance of foam sheathing products range from 3.85 per inch to 7.30 per inch depending on the type of product used.

5. Provides an uninterrupted envelope of insulation—The sheathing application also insulates the framing members of the structure, thereby reducing the thermal shorts created by those members.

6. Common construction is not altered significantly in order to increase thermal performance—Typical 2" x 4" stud (16" on center) walls are the most common wall construction and its thermal performance is improved with little or no alteration of the framing (see Disadvantage 2).

7. Partial offset of cost—Being a sheathing material, the investment of plastic foam sheathing is at least partially offset by the cost of the conventional sheathing that is replaced.

Disadvantages

1. Some breakage problems—Some foam sheathing products tend to break more easily than conventional sheathing products and must be handled with appropriate care. The higher density, thicker, or faced products tend to exhibit less breakage problems.

2. Nonstructural sheathing—Foam sheathing products are nonstructural and like other nonstructural sheathings cannot be used as a nail base and are required to be augmented with acceptable corner bracing.

3. Some finishing construction alterations are necessary—For those products thicker than ⅝", it is usually necessary to utilize longer nails and extended window and door jambs.

4. Interior finish restrictions—Being combustible, it is required by code that the interior be covered with at least ½" thickness of gypsum board.

Interior Wall Substrate Assemblies

Rigid plastic foam boards can also be applied on the interior of the frame wall behind the gypsum drywall. There would be no difference in the calculated thermal resistance (R-value) as compared with exterior applications. However, there are some disadvantages associated with interior applications. The framing members of the wall are closer to outside temperatures; sheathing is still needed; under transient temperature cycles, the insulation is less effective on the warm side of the wall composite; the living space is reduced under the same exterior dimension. The advantage is that certain foams on the warm side can serve as a vapor barrier.

Rigid plastic foam boards can be economically applied to the interior walls of residential masonry construction. The predominant application of this assembly is in industrial/commercial buildings, but the application for residential construction is identical.

Masonry Wall Construction

Solid masonry wall construction constitutes a small fraction of residential wall construction, with primary usage being for commercial/industrial buildings.

RESIDENTIAL FLOOR/FOUNDATION ASSEMBLIES

Table 4.14 describes the predominant types of floors and foundations found in residential buildings. The primary foundation construction is masonry. The primary type of foundation is slab-on-grade, followed closely by full/partial basement. Floors other than slab-on-grade are framed primarily via wood joist/beam, 2" x 8" spaced 16" on center.

Table 4.14: Floor/Foundation Components (1)

| | . . . . .Region . . . . | | | | |
	North-east	North Central	South	West	US Total
Foundation construction					
Concrete block	28.8	45.0	33.8	2.4	28.7
Poured concrete	67.8	51.5	60.7	95.6	67.3
Other and no answer	3.4	3.5	5.5	2.0	4.0
Total	100.0	100.0	100.0	100.0	100.0
Floor framing					
Wood joist/beam	89.9	88.6	45.6	59.9	61.2
Other and no answer	10.0	2.4	4.5	8.2	5.6
Not applicable (concrete)	0.0	9.0	49.9	31.9	33.3
Total	99.9	100.0	100.0	100.0	100.1
Foundation type					
Full/partial basement	51.8	71.4	20.4	24.7	36.8
Crawl space	3.0	7.0	16.5	25.4	15.0
Slab-on-grade	43.3	18.0	60.0	42.4	44.3
Other and no answer	1.9	3.5	3.1	7.4	4.0
Total	100.0	99.9	100.0	99.9	100.1

Source: BNL-50862

Floor/foundation perimeter insulation applications utilizing plastic foams have been used for many years on industrial/commercial buildings. However, some residential applications are slightly different. Applications shown in Figure 4.6 are for rigid plastic foams. Some foams should have a moisture barrier, since they are more susceptible to moisture pickup and, consequently, are also subject to deterioration from freeze-thaw cycling. The advantage of this type of assembly are ease of application, continuity of the exterior building insulation envelope, and retention of the interior mass to dampen the interior temperature cycling. The primary disadvantage is that the section between the siding and the grade must be covered to prevent UV deterioration, damage, and exposure, to possible sources of ignition.

Floors over crawl spaces may be insulated with mineral fiber batts either by insulating the foundation walls if the crawl space is unvented, or if the crawl space is vented and unheated by placing insulation between the floor joists. The former method, as described below, is usually more economical. The vapor barrier (polyethylene film) covering the crawl space ground is placed against the wall (Figure 4.7a), and then the insulation is installed. One edge of insulation is placed on top of the foundation wall with the remainder of the insulation draped over and against the inside of the wall. The main disadvantage of this application is that the sill is not fully insulated.

Floors over vented crawl spaces (Figure 4.7b) can be insulated by installing insulation between the joists. The insulation should be pushed tightly against the subflooring above and is held in place with stiff wire fasteners or flexible wire laced under the joists. In all cases, the vapor barrier side of the insulation should face the floor above—that is, be adjacent to the warm side in winter. Polyethylene may be used as a vapor barrier to cover the ground. If it is desired to protect the insulation from the weather, such as may be the case in open crawl spaces, nail standard, interior grade softwood plywood, nail-base insulation board, or similar covering to the bottom of floor joists.

In some cases, blanket insulation is located between furring strips on the interior wall (usually unit masonry and/or poured concrete) when finished rooms are located in the basement. Loose fills, such as perlite, vermiculite, or plastic foam beads, may be installed in the cavities of unit masonry to increase thermal resistance.

Figure 4.6: Rigid Plastic Foams for Floor/Foundation Perimeters

Slab-on-grade

a.

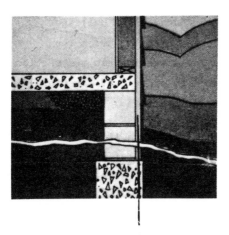

Rigid cellular
plastic foam

(continued)

Figure 4.6: (continued)

Basement

b.

Rigid cellular
plastic foam

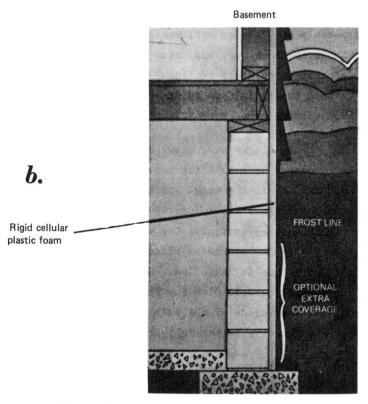

Source: BNL-50862

Figure 4.7: Insulation for Floors over Crawl Space

a.

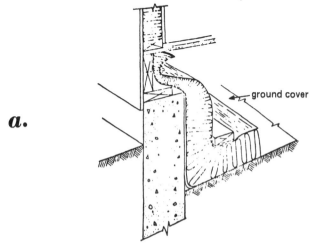

(continued)

Figure 4.7: (continued)

b.

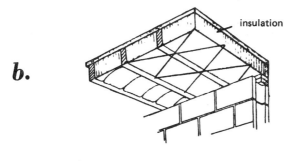

insulation

(a)　Vented
(b)　Unvented

Source:　BNL-50862

PREFABRICATED WOOD FRAMED STRUCTURES

Most manufacturers of prefabricated wood framed dwellings supply R-11 or R-13 mineral fiber batts for walls although an alternate of 6" stud walls with R-19 batts can be used. Ceilings available are R-22 to R-40 batts or R-30 and R-38 blown cellulose insulation. Floors over unheated crawl spaces are R-11 to R-19 batts. Installed cost figures are approximately the same as those for site-built structures.

Energy savings packages featuring increased insulation, storm or multiple-glazed windows, and thermal doors are available. The return on investment for the energy savings packages was calculated to be 3 to 5 years for 2" x 4" stud walls and 12 years for the 2" x 6" stud walls. Also used can be air infiltration control in the energy package with sealing of joints, electrical boxes, etc.

RESIDENTIAL INSULATION INSTALLED COSTS

A survey of thermal insulation installation contractors who install insulation, mainly in wood frame, newly constructed dwellings, as specified by the general contractors or owners, gave the following results. Only batt or blanket insulation was reported to be used in wall cavities, with a thermal resistance of R-11 as the most common. Some contractors reported that up to 10% of their installations involved 6" stud walls with R-19 batts installed. Others reported use of R-11 or R-13 batts with plastic foam sheathing used in place of plywood or wood fiberboard sheathing. These both provide a total wall resistance of up to R-21 or a U as low as 0.05.

There are debit factors. 6" studs are less available and more expensive than 4" studs and the use of 6" studs decreases the enclosed areas by 2" for each wall, where equal outside dimensions are assumed. For a 47' x 33' single story house, the decrease in living area is 27 ft^2 in approximately 1,240 ft^2. Plastic foam sheathing does not provide racking strength to the walls so the added expense of wall bracing must be included.

Ceilings are most often insulated with blown or blanket-type insulation with cellulose, fiber glass, or rock wool as the common types. Cellulose is, of course, only installed by blowing. A thermal resistance of R-19 was most common with some contractors reporting installations up to R-30. Demand for resistance over R-30 was found small.

Floors over unheated spaces are commonly insulated with batts, usually supported by wire laced over the bottom face of the joists. A thermal resistance of R-19 was again the most common. Floor slabs were commonly insulated with 24" widths of ¾" or thicker plastic foam placed around the perimeter. In some instances, blanket or foam insulation is placed on the interior face of basement walls, but the above are most commonly installed by other contractors.

A rough rule of thumb for 1977 installed cost of batt insulation was a base price for kraft paper-faced batts of 1 to 3 cents per ft² above the R-value, i.e., an R-11 batt would cost between 12 to 14 cents per ft². This corresponds to about 1 cent per ft² per R. Foil-faced batts command a premium of about 1 cent per ft², while unfaced batts with a 2 or 4 mil polyethylene vapor barrier installed separately require a premium of about 2 to 4 cents per ft². Ceiling installation of batts by an experienced contractor costs about the same as wall installation as the workers commonly operate on stilts for 8' ceilings. Higher ceilings cost more with the price increasing with height and the staging required. Floors over crawl spaces may cost the same with cost increasing with more difficult working conditions.

Blown insulation is priced in a similar manner. There is a need for installation of air deflectors near the eaves to prevent blockage of ventilation for moisture control and care in not placing insulation near sources of heat. Insulation can provide the greatest reduction in heat loss when located in attics. Installation of mineral fiber batts or loose fill insulation for this application is less expensive than for walls. Therefore, the attic should be one of the first places to examine for upgrading thermal performance.

Floors in need of retrofitting are more expensive. This is, again, due to obstructions where the crawl space has been used for storage, limited working space, and generally difficult working conditions. Again, the premium varies with conditions.

Refitting walls is considerably more difficult. In the cases where the interior wall surface is to be replaced, the problem is simple as batts may be installed as for new construction. Otherwise, insulation must be blown into each stud space by removal or cutting a hole in the exterior or interior facing and repairing the hole. Obstructions such as fire stops, plumbing, electrical assemblies, or ducts require careful attention to obtain full insulation value. Mineral wool or cellulose blowing insulation, or urea-based foam-in-place insulation may be installed. Cost will be higher and efficiency questionable due to internal obstructions.

Improvement may also be obtained by exterior installation of foamed plastic sheathing or foam or other panels installed on the interior. Both foam board

Table 4.15: Major Insulation Usage in Residential Construction

	Mineral Wool Batts	Mineral Wool Loose	Cellulose Loose	Wood Fiberboard	Plastic Foam Board	Urea-Formaldehyde Foam-in-Place
Roof/ceiling						
In-frame cavities	N–R	N–R	N–R	—	—	—
Above roof sheathing	—	—	—	N	N–R	—
Cathedral ceilings	N–R	—	—	N–R	N–R	—
Walls						
In-frame cavities	N	R	R	N	—	R
Sheathing	—	—	—	N	N–R	—
Floors						
Wood joisted	N–R	—	—	—	—	—
Concrete slab	—	—	—	—	N	—
Basement wall						
Exterior	—	—	—	—	N	—
Interior	N–R	—	—	—	N–R	—
Approximate cost ($/ft²/R)	0.01	0.01	0.01	0.04–0.05	0.02–0.05	0.04–0.06
Advantages	low cost, non-combustible without facings		low cost availability	may provide racking support and nail holding properties as sheathing	high R/inch, may act as infiltration seal, low water absorption	high R/inch, may act as infiltration seal
Disadvantages	facings may be combustible		combustible	combustible	combustible, non-structural	combustible, foam may shrink in place which reduces the effectiveness of the insulation and may permit air infiltration

Note:
N — used in new construction.
R — used in retrofitting.

Source: BNL-50862

methods usually require additional trim at window or door openings due to the increased thickness.

SUMMARY

The most commonly used residential insulations, where they may be used for new construction and retrofit applications, and some advantages and disadvantages of each are presented in Table 4.15.

REFERENCES

(1) "Survey of Single Family Home Characteristics," National Association of Home Builders (1974).
(2) F.W. Dodge Division, McGraw-Hill (December 1976).
(3) Survey Report: "U.S. Residential Insulation Industry," Office of Business Research and Analysis, U.S. Department of Commerce (August 1977).
(4) Jones and Hendrix, "Residential Energy Requirements and Opportunities for Energy Conservation," Center for Energy Studies, Univ. of Texas (September 1975).
(5) "The Arkansas Story," Report No. 1, Energy Conservation Ideas to Build On, Owens-Corning Fiberglas Corporation (4-BL-6958A) (1975).
(6) "Insulation Manual," National Association of Home Builders, Research Foundation, Inc.
(7) Rossiter, W.J., Jr., Mathey, R.G., Burch, D.M., and Pierce, E.T., NBS Technical Note 946 (July 1977).
(8) Housing and Urban Development, "Usage of Materials," (1974).
(9) National Mineral Wool Insulation Association, Inc., "How to Insulate Homes for Electric Heating and Air Conditioning" (1977).

INDUSTRIAL/COMMERCIAL BUILDING INSULATION ASSEMBLIES

The material in this chapter was excerpted from a report prepared by Brookhaven National Laboratory with the assistance of Dynatech R/D Company (BNL-50862).

There are many building assemblies used in industrial/commercial building construction. This discussion is limited to the typical assemblies shown in Sweets Catalog (1). In Table 5.1 is presented a tabulation of the most commonly used commercial/industrial insulations, where they are applied, and their advantages and disadvantages.

ROOF/CEILING INSULATION ASSEMBLIES

Insulation of roofs is the most efficient insulation operation in commercial/industrial buildings. The roof deck is usually flat and may be constructed of cast masonry, preformed masonry planks, steel planks or panels, or wood. In practically all instances, the insulation is located above the roof deck structure, and a large majority have the insulation covered by the built-up roof. This is the conventional overdeck assembly, Figure 5.1. Other techniques for roof insulation assemblies include insulation roof membrane assemblies, underdeck assemblies, and structural panel assemblies.

The structural panel assemblies described in the previous chapter for wall systems are also applicable to industrial/commercial roof/ceiling assemblies. Ceiling truss/rafter cavity assemblies and ceiling substrate assemblies are rarely used for industrial/commercial applications. A description of these last two designs was presented in the last chapter describing roof/ceiling assemblies for residential buildings.

Conventional Overdeck Assembly

The conventional overdeck assembly depicted in Figure 5.1 uses insulation between the roof deck and roof covering. The insulation is usually provided in

Table 5.1: Major Insulation Usage in Commercial/Industrial Structures

	Mineral Wool Batts	Wood Fiber Board	Mineral Board	Plastic Foam Board	Plastic Foam-in-Place	Combined Mineral-Foam Boards	Insulating Concrete Air Entrained	Insulating Concrete Lightweight Aggregate
Roof/Ceiling								
Above roof deck	—	n-r	n-r	n-r	n-r	n	n-r	n-r
Below roof deck	n-r	—	—	—	r	—	—	—
Walls								
In cavities	n-r	—	—	n	n-r	—	—	—
Sheathing or siding	—	—	n-r	n-r	—	—	—	—
Floors								
Concrete slab	—	—	—	n	—	—	n-r	n-r
Wood or steel joists	n-r	—	—	n-r	n-r	—	n-r	n-r
Thermal Resistance (per inch thickness)	2.9-4.6	2.8	2.8	5.0-7.7	5.0-7.7	4.5-7.3	......0.3-1.8*.......	
Approximate Cost ($ per R per ft^2)	0.01	0.04-0.05	0.05-0.06	0.03-0.05	—	0.04-0.05	—	—
Advantages	low cost, non-combustible	availability	limited combustibility	high R/inch	high R/inch	mineral board acts as a fire barrier to protect foam	non-combustible	non-combustible
Disadvantages	facings may be combustible	combustible	less thermal efficiency than foam or combined board	must be protected from fire exposure as it is combustible		greater thickness per R than foam alone; foam is combustible	greater thickness per R than other products	

*Depending on density.

Note: n = used in new construction; r = used in retrofitting.

Source: BNL-50862

board form, ranging from 2 x 4 ft to 4 x 8 ft panels. Materials in use include wood fiberboard, perlite board, fiber glass board, polystyrene, urethane, or isocyanurate foam boards, or composite foam and mineral boards. The purpose of the combined mineral and foam boards, where the mineral board is installed next to the roof deck, is to provide a fire barrier to prevent flame penetration of the roof from interior or exterior fires and spread of fire below the roof due to combustibles penetrating the roof and adding fuel to an interior fire.

Figure 5.1: Conventional Overdeck Assembly

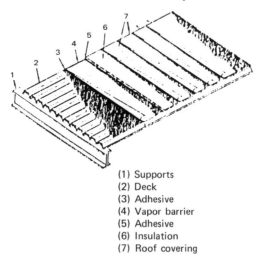

(1) Supports
(2) Deck
(3) Adhesive
(4) Vapor barrier
(5) Adhesive
(6) Insulation
(7) Roof covering

Source: BNL-50862

Limitation of fuel contribution was developed following the 1953 large, undivided Livonia automobile plant fire where roofing asphalt penetrated steel roof plant and caused extensive spread of interior fire. Combustible roof insulation was found adequate with limited quantities of approved adhesives to reduce spread of flame. Roof insulation of limited combustibility can attain hourly fire resistance ratings (¾ hour and greater).

More recently, some foam plastic has qualified for use directly over steel or other decks without use of the low combustibility protective layer. It still appears necessary to use protective barriers such as perlite or gypsum board to attain hourly fire resistance ratings. Attaining Class A, B, or C ratings for exterior fire exposure depends on the exterior surfacing material.

Other insulation materials include insulating concrete and plastic foams of urethane or isocyanurate. The cementitious materials include perlite concrete, vermiculite concrete, and air entrained concrete. They are available in densities as low as 25 pcf. Densities below 100 pcf are not recommended as load bearing decks, but may be used as insulating fill layers above structural decks. The thermal conductivity increases with density. Poured gypsum concrete is available with wood chip or perlite aggregate. The former has an advertised R of 0.67 per inch, while the latter has an R of 0.87 per inch. The product with wood chip aggregate may not attain a classification of noncombustible. Values of thermal conductivity similar to the above may be attained with lightweight aggregate concrete. Aggregates include perlite, shale, vermiculite, etc.

Roof insulation, with few penetrations to serve as heat flow short circuits, makes a textbook case for calculation of total thermal resistance or thermal transmission. The manufacturer must provide adequate information on the product sold and it is then a matter of using values compiled by ASHRAE for resistance of the built-up roof, the roof deck, and the inside and outside air film resistances.

All manufacturers provide data on thermal conductance, as required by several Federal Specifications, or thermal resistance. Some provide thermal conductivity data. Several manufacturers provide examples of heat transmission calculations or "U" values for various configurations of metal, gypsum, concrete, or wood decks for winter and summer conditions as well as energy conservation calculations based on various locations and energy costs.

Most of those calculations are based on values obtained at 75°F mean temperature. A more precise practice would be to provide data at mean temperatures consistent with winter or summer conditions. For 70°F inside conditions, a winter mean temperature of 40°F and a summer mean temperature of 90°F are suggested. This is equivalent to outdoor temperatures of 10°F for winter and 110°F for summer. These appear adequate.

Insulated Roof Membrane Assemblies

This assembly differs from the conventional overdeck assembly in the relative location of the insulation. For the conventional system, the insulation is between the deck and membrane, whereas this assembly places the insulation above both the deck and membrane as shown in Figure 5.2. Extruded polystyrene foam is suited for this application because of its high moisture resistance.

Figure 5.2: Insulated Roof Membrane

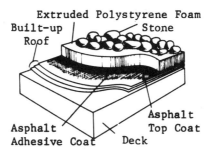

Source: BNL-50862

The design concept of the insulated roof membrane assembly is essentially the same for all roof decks. Complete instructions and design sheets are supplied by the manufacturer. Basically, the design concept is to put the built-up roof membrane directly on the wood deck, apply boards of extruded polystyrene to the asphalt top coat, and finally, cover the insulation with a specified layer of stone. The advantages include improved membrane life, ease of application, and ease of adding additional insulation later, if desired.

The insulated roof membrane assembly may be used for reroofing. If the old roof must be totally removed because of deterioration, the insulated roof membrane assembly can be placed on the new membrane just as in new construction. Where the roof membrane remains essentially intact, some patching and one or two new plies will be sufficient. An asphalt top coat, the desired thickness of insulation and the specified stone cover complete the retrofit assembly. The most important precaution is to make sure that the old roof assembly can accommodate the load of stone.

Underdeck Assemblies

In underdeck assemblies, the insulation is applied below the roof deck. Mineral fiber batts or board insulation are often used in the frame cavity, being held up by wiring or interior finishing. Mineral fiber blankets are often draped over metal roof purlins prior to application of the metal roof panels on many prefabricated metal buildings. Spray-on-foam or cellulose insulation could be used, but this would require sufficient interior finishing to meet fire resistance standards.

Prefabricated Metal Buildings

Many metal buildings are produced with corrugated or shaped metal facings as the roof surface. These must be insulated below that metal facing. Fiber glass blankets are the major insulation material used for this application. Plastic foam boards are also used in conjunction with fiber glass blankets. The foam boards are laid across the metal purlins or trusses and the metal roof surface is then placed over the foam boards and attached. This method of assembly reduces the thermal shorts caused by the metal structural members. The fire rating of these materials must be considered in any design.

INDUSTRIAL/COMMERCIAL WALL ASSEMBLIES

The basic methods of wall construction in industrial/commercial buildings are as follows:

Steel Frame — Welded/riveted load-bearing beam construction. Typically uses sandwich panel/curtain wall insulation and insulated frame cavity assemblies. Applicable to high rise and low rise buildings.

Reinforced Concrete — Poured and reinforced concrete. May use sandwich panel, cavity wall, interior wall substrate, or stucco base insulating assemblies. Used for low rise buildings.

Load Bearing — Concrete block and/or brick construction. May be sandwich panel, cavity wall, interior wall or stucco base insulating assemblies. Mostly used for low rise buildings.

Prefabricated — Prefabricated metal buildings typically use sandwich panel/curtain wall, insulated frame-sheathing and insulated blanket frame-cavity assemblies. Mostly used for low rise buildings.

Combinations — Combinations of the basic walls listed above are common. The insulating assemblies match those listed above.

The major assemblies reported in manufacturers' literature and Sweets Catalog are:

(1) Cavity wall assemblies,
(2) Interior wall substrate,
(3) Insulated frame construction,
(4) Stucco base assemblies, and
(5) Sandwich panel/curtain wall assemblies.

Cavity Wall Assemblies

Cavity wall assemblies basically consist of attaching rigid board insulation to the outside of the structural masonry wall before adding the exterior veneer brick. Illustrations of typical systems are shown in Figure 5.3. The size of the cavity between the masonry wall and the brick veneer is limited by standard construction practices. Therefore, it is usually advantageous to use insulation materials with a high R-value per unit thickness in order to obtain reasonable thermal performance. Rigid polystyrene and polyurethane foam board stock are the major insulations used in this application.

The insulation boards are normally installed with the long dimension horizontal and they are attached to the load-bearing masonry wall with some type of mastic adhesive. Brick ties are provided to tie the exterior face brick to the load-bearing masonry wall. Normally a ½-inch to ¾-inch air space is provided between the insulation board and the outer wythe of brick to provide finger space for placement of brick and to provide a drainage passage for wind-driven rain which may penetrate the face brick.

Weep holes are provided at the bottom of the wall to drain this excess moisture. Vapor barriers may or may not be recommended in this system depending on the sensitivity of the insulation to moisture and the inherent water vapor transmission properties of the material itself.

Another variation on cavity wall construction is the use of loose fill materials such as loose-expanded polystyrene beads, reprocessed foam granules, perlite, or vermiculite as a pourable masonry fill insulation. This material is poured within the cavities of masonry wall construction. Normal precautions for wind-driven moisture must be followed. Perlite loose fill treated with silicone for moisture resistance is accepted and available.

Both the board and loose-fill applications find their widest use in new construction. The systems are not limited to exterior veneers using stackable wythes of brick but also could be used with granite, slate or masonry veneer panels as exterior cladding. However, the applicability for retrofit is limited.

Figure 5.3: Cavity Wall Assemblies

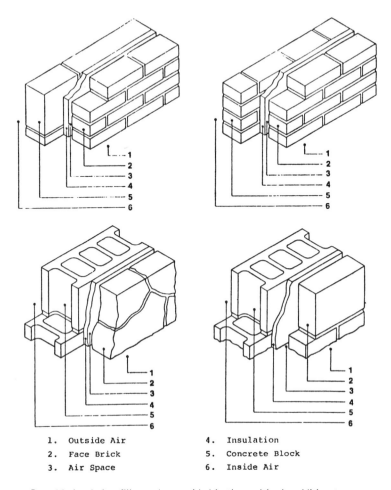

1. Outside Air
2. Face Brick
3. Air Space
4. Insulation
5. Concrete Block
6. Inside Air

Pourable insulation fill may be provided in the cavities in addition to or instead of the board insulation.

Source: BNL-50862

It is not likely the loose-fill system would be used as retrofit at all, while the board system would only be economical if an exterior "face-lift" were included as part of the project plans. If energy conservation were the only motive for building modification, the cost of the exterior finish would cause the project to have excessively long payback periods (i.e., greater than 15 years).

The thermal performance of cavity wall assemblies is directly dependent on the type and thickness of insulation used. There is an inherent advantage in the use of insulation on the outside of the masonry wall insofar as the main mass of the structure is insulated against the extremes of the annual and daily temperature cycle. The amplitude of temperature changes "seen" by the structural masonry wall is damped by the insulation and, therefore, remains closer to the conditioned space temperature. Theoretically, this will allow the HVAC equipment to operate more efficiently without having to compensate for wide temperature changes. However, a meaningful analysis of this benefit is outside the scope of desk-type calculation and requires the use of a sophisticated computer program.

The cavity wall assemblies in common use today make use of polystyrene or polyurethane board insulation approximately one inch thick, and this type of wall has U-values between 0.10 and 0.15 as compared with a noninsulated wall U-value of approximately 0.30.

Because of the masonry components in the cavity wall system, the walls have inherently high resistance to fire. Building codes prescribe the allowable flame spread and smoke values for the insulating materials used in these types of walls. Fire-stopping is required in the cavity of each floor level of the building. Because of the encapsulation of the insulation between the two masonry surfaces, the safety threat to occupants in the event of a fire is remote. There are no requirements or standards with respect to toxicity and odor for these cavity wall assemblies nor is there need for any.

The significant advantage of the cavity wall systems is the inherently high fire safety associated with these systems. The disadvantage is the limited ability to upgrade thermal resistance after construction through retrofitting.

Interior Wall Substrate Assemblies

Interior wall substrate assemblies involve attaching a layer of rigid plastic foam, mineral fiber, or other insulation to the interior surface of the load-bearing masonry wall and then adding an interior finish material, typically ½-inch thick gypsum board. Illustrations of typical constructions using rigid plastic foam board are shown in Figure 5.4.

When rigid plastic foam board insulation is used in this assembly, wood furring strips may be provided for attachment of the gypsum board interior finish. When furring strips are not used, mastic-type adhesives are used to bond the gypsum board to the insulation. Normally wood-nailer strips are provided at the top and bottom of the wall to provide some mechanical attachment for the gypsum board. Also, wood-nailer strips are provided at locations where heavy wall fixtures such as cabinets are to be hung. Electrical runs must either be surface mounted or provide for shallow receptacle boxes (1.5").

When fiber glass blankets are used it is necessary to attach wood furring strips or to build an interior stud wall assembly on which to attach the interior finish material. Fiber glass blankets do not provide any support for the interior finish and require greater thicknesses to provide insulation values equivalent to the plastic foam systems.

The gypsum dry-wall systems are normally used in new construction; however, they do have retrofit applicability. A masonry substrate is not required for the insulating layer; it could be applied over any sound existing wall surface. This assembly could be one of the least expensive retrofit systems, particularly if some remodeling is included in the building retrofit plans. Minor trim modifications would be required around window and door openings and around electrical outlets. Normally these details would be no great burden in a modification of this sort.

Figure 5.4: Interior Wall Substrate Assemblies

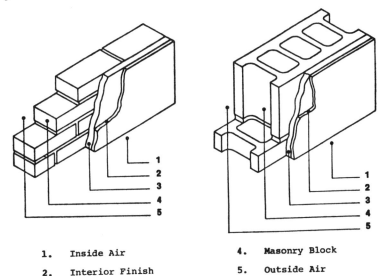

1.	Inside Air	4.	Masonry Block
2.	Interior Finish	5.	Outside Air
3.	Rigid Plastic Foam		

Source: BNL-50862

As in the cavity wall assembly, the thermal performance will be dependent on the type and thickness of insulation used. No air infiltration benefits are expected and heat loss is calculated by straight conduction methods. Since the insulation is applied to the conditioned space side of the wall, no benefit from the thermal mass of the building can be claimed. Again, it should be pointed out the thermal-mass benefit has not been quantified to the point where it is accounted for in standard heat loss calculations.

Typically 1 to 2 inches of plastic foam insulation is used in this application, which results in overall thermal transmittance values (U) of from 0.08 to 0.15 Btu/h ft^2 °F compared to noninsulated walls which have a transmittance of from 0.30 to 0.40. Systems utilizing fiber glass blankets will have U-values of approximately 0.20 at comparable thicknesses.

The resistance to fire of this particular system is dependent on the type of insulation used and in some cases on the thickness of the interior finish provided. All model building codes allow the use of plastic foams, provided the plastic foam is fully protected from the interior of the building by a thermal barrier of ½-inch gypsum board having a finish rating of not less than 15 minutes by ASTM E119 test or other approved materials having an equivalent finish rating. Greater endurance ratings can be obtained by using thicker layers of Type X gypsum board. The manufacturer's literature for all plastic foam insulation makes note of the 15-minute building code requirement. Standard nailing gypsum board is required to meet this 15-minute requirement. As with other applications, there are no requirements or data with respect to toxicity or odor.

The advantages of this wall system with foam board insulation include: (1) ease of applicability for retrofit, (2) low labor intensity if furring strips are excluded, (3) a completely insulated wall without thermal shorts, and (4) a structurally-sound wall since the interior finish is completely supported by the insulating layer. The principal disadvantage is the loss of the "thermal mass" effect.

Insulated Frame Construction

Practically all residential construction and a large portion of industrial/commercial construction are the insulated frame assembly with mineral fiber insulation commonly used in the frame cavity. This was discussed for residential construction in the previous chapter.

One popular system for improving the thermal efficiency of industrial/commercial walls is to use plastic foam insulation as exterior sheathing in combination with mineral fiber insulation in the cavity as illustrated in Figure 5.5. In this system the insulating capacity of the wall is economically maximized by using the lower cost per unit volume material (i.e., mineral fiber) in the available cavity space, and using the highest R per unit thickness material (i.e., plastic foam) in the area where space is limited.

Not only does the plastic foam sheathing add to the thermal resistance across the cavity, but it insulates the thermal short circuits caused by the framing members. The efficiency of the mineral fiber insulation is improved because the reduced ΔT across the cavity results in reduction of any remaining convection currents. In addition, air infiltration to the interior is reduced.

An alternate assembly is to spray-in-place a rigid polyurethane type foam insulation in the cavity. This foaming-in-place technique takes place from the inside of the structure after the exterior sheathing or siding is in place. Advantages of this type of system are the complete sealing of all openings with foam and a very high R-value per unit thickness. However, a vapor barrier and a thermal fire barrier with at least a 15-minute finish rating need to be added to the interior surface.

Another approach is to insert preformed blocks of plastic foam insulation in the cavity. This type of application is not in common usage and would only be justified economically where the absolute minimum U-value is required for a fixed wall thickness.

With respect to installation methods, the plastic foam sheathing system is applied in the same manner as any other type of sheathing material. The plastic foam is not a structural sheathing. Therefore, bracing of some form is required between the framing members at the corners of the building to meet rack loading requirements (ASTM E-72 Tests). Also, the plastic foam cannot act as a nail base so exterior siding materials must be attached to the framing members.

Figure 5.5: Insulated Frame Construction

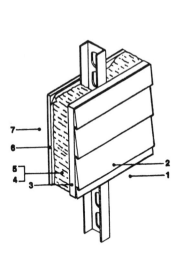

1.	Outside Air	5.	Mineral Fiber Insulation
2.	Exterior Siding	6.	Interior Finish
3.	Plastic Foam Insulation	7.	Inside Air
4.	Cavity		

Source: BNL-50862

Installing spray-in-place polyurethane requires specialized equipment and trained operators which are provided by the insulation supplier. Proper safety precautions with respect to personnel exposure to component chemicals are important and must be followed to eliminate hazards.

The preformed block insulation for cavity fill requires no specialized installation techniques other than careful precutting of the block to assure an accurate fit in the cavity.

All of the above methods can be used for new construction. Only the plastic foam sheathing system has real application for retrofit. Several board manufacturers have developed systems in which the plastic foam sheets are attached to the exterior of the wall over the old finish material and a new finish is added to cover the foam. This system is particularly applicable to wood-frame-type construction. Metal framing would present a fastening problem unless the existing siding can act as a nail base or mechanical fasteners or adhesives are used.

The thermal performance of insulated frame construction will be dependent on the type and thickness of insulation used. The plastic foam sheathing assembly will tend to deliver the best cost-performance index when the system includes mineral fiber insulation in the cavity. No manufacturer lists actual performance data for nonresidential buildings using any of the three systems discussed.

This should come as no surprise, however, since it is impossible to select a "typical" nonresidential building. All plastic foam manufacturers do list overall U-values for various wall systems when their product is used and the range normally is 0.05 to 0.08. These values take no credit for energy savings via reduced air infiltration and this savings can be substantial (>10%) depending on the type of foam used and the type of construction.

All building codes will require an interior finish material with a 15-minute fire resistance rating if plastic foam is used as part of the wall construction. All plastic foam manufacturers point out this requirement in their literature and will normally include ASTM E-84 Test data for flame spread and smoke developed for their product.

Stucco Base Assemblies

Stucco base wall assemblies could be included as merely another exterior cladding assembly for either frame wall or masonry wall construction. However, several plastic foam manufacturers list the stucco type of construction as a separate category in their literature and point out unique benefits of this system. Two manufacturers have developed a modified thin veneer stucco finish for use over molded polystyrene foam. Other manufacturers of rigid plastic foam issue detailed specifications for conventional three coat stucco finishes.

In the latter case, the installation practice is identical to normal stucco application with the exception that the rigid foam boards replace the back-up paper on frame construction or in masonry construction; the boards are attached to the wall prior to installing the reinforcing mesh.

The two thin veneer systems differ substantially from conventional stucco. In both cases, a fiber glass loose-weave fabric forms the reinforcement and only two coats of a specially formulated cementitious coating are applied. This results in a surface coating of approximately ⅛ inch thickness as opposed to the conventional stucco thickness of one inch. Although the main thrust in the promotion of this system is in new construction, the two thin veneer systems are suitable for retrofit applications when an exterior face-lift is desired, particularly in block wall construction where the original wall was noninsulated and has no exterior finish except paint.

The thermal performance of any assembly is dependent on the thickness of the foam layer. When applied to an 8-inch concrete block wall the thin veneer systems have U-values ranging from 0.09 for 2-inch thick foam to 0.13 for 1-inch thick foam. The noninsulated concrete block wall would have a U-value of 0.29.

Masonry wall assemblies are highly fire resistive and the inclusion of plastic foam insulation as part of a stucco finish system does not significantly change the fire performance. With frame wall constructions, all manufacturers point out the need for interior thermal barriers in the design. When the systems are properly designed, the fact that they incorporate plastic foam does not increase the fire hazard over commonly accepted alternate systems.

The prime advantage of a stucco base system is the ease with which it can be incorporated into retrofitting masonry block walls. Masonry block walls cannot accommodate some of the other popular retrofit insulation types such as urea-formaldehyde foam or blown cellulose, and therefore, this type of system could fill a significant need in the nonresidential building category.

One limitation which still exists with the proprietary thin stucco veneer systems is the longevity of the surface coat. Because of the dissimilarity in coefficients of thermal expansion between the cementitious surface coat and the cellular plastic, questions arise concerning cracking and delamination from thermal cycling during the life of a building. The thicker stucco coatings, because of their reinforcement, larger thickness and history of performance, are not as suspect of this type of failure.

Sandwich Panel/Curtain Wall Assemblies

Sandwich panels are a building component in which rigid plastic foam may be foamed-in-place between two rigid facings, or foam or other board stock may be adhesively bonded to two facings to form a monolithic panel. This type of panel offers both high insulating value and structural strength. Since these panels are commonly prefabricated in a factory, labor savings are realized at the construction site due to faster assembly. Commonly, corrugated steel and aluminum (0.32-inch thick) are used as facings, but plywood, gypsum board, or polymer-modified concrete may also be used. Thickness of the panels normally ranges from 1 to 2 inches, although panels up to 6 inches thick may be used in low temperature space applications.

Sandwich panels fit a variety of construction applications. Full height panels may be placed over single-story steel-frame construction to form the complete walls or roof. They may also be used as curtain walls or spandrel panels in multistory steel frame buildings.

In retrofit applications, they may be used to replace windows or portions of windows. They can be attached as an overlay for masonry walls. Occasionally, panels may be used as the walls of single-story buildings without additional framing, in which case the panel forms the bearing wall. For retrofit applications over an existing wall, a single-faced panel (linear panel) is commonly used. Commonly, rigid board insulation is used if the panels are to be adhesively laminated and polyurethane type foams are used for the foam-in-place type panel. Manufacturers of these panels promote both new construction and retrofit applications about equally.

The thermal performance of the systems is dependent on thickness and material composition. A typical 1-inch panel with 26-gauge steel facings will have a U-value of 0.23 with a molded bead polystyrene foam core, a U-value of 0.20

with an extruded polystyrene foam core, a U-value of 0.16 with a urethane core, and a U-value of 0.33 with a fiber glass core. Systems of this sort do have the advantage of insulating the thermal short circuits caused by the steel framing members.

When one surface of a foam panel forms the interior surface of the building, all building codes require that either a 15-minute thermal barrier be provided or, when sprinklers are provided, the facing material be at least 28-gauge steel or 32-mil aluminum. Some insurance companies will require testing and approval by Factory Mutual Corporation large wall-corner test. Several manufacturers' panels carry this approval or listing. None of the approvals carry requirements with respect to toxicity and odor other than FDA approval of the facing material used on the interior of food processing rooms. Significant advantages of the panel systems are:

(1)   High thermal resistance per unit thickness,
(2)   High strength to weight ratio for wall systems,
(3)   Thermal properties not degraded by exposure to moisture,
(4)   Lower labor intensity during construction, and
(5)   Insulation of thermal short circuits through framing members.

Limitations associated with this type of construction system are:

(1)   Poor acoustics from low mass systems,
(2)   Limited flexibility in architectural design,
(3)   Difficulty in incorporating electrical and water lines in the wall, and
(4)   Limited fire resistance characteristics.

Other faced panel systems use mineral fiber insulation in the voids with or without insulation boards placed at some inside location. In many cases, the metal faces penetrate the assembly for strength. Such penetrations can form a thermal short circuit, but insertion of plastic "thermal breaks" can minimize such short circuits.

Fire resistive faced panel systems are available. These commonly incorporate mineral panel insertions or additions, such as gypsum board, and are available in ¾- to 4-hour time temperature ratings.

FOUNDATION/FLOOR INSULATION ASSEMBLIES

In foundation insulation applications, also known as perimeter applications, rigid plastic foams are widely used. In this application, the insulation is in constant contact with the earth and is subjected to back fill pressure, active ground chemicals which may be under pressure, and active biological mechanisms. Due to water resistance and closed-cell structure, some rigid plastic foams can withstand this exposure and continue to maintain insulating effectiveness for the building life.

Sketches of typical construction details are shown in Figure 5.6. When the floor slabs carry heat ducts or resistance heating cables, the insulation may be continuous beneath the floor slab. In most cases, however, the insulation is applied to the vertical face of the foundation wall. All types of plastic foams are used in this application including the spray-in-place systems. Some manufacturers of molded polystyrene and urethane board stock recommend vapor barriers be applied to the side of the insulation in contact with the earth.

Figure 5.6: Floor/Foundation Assemblies

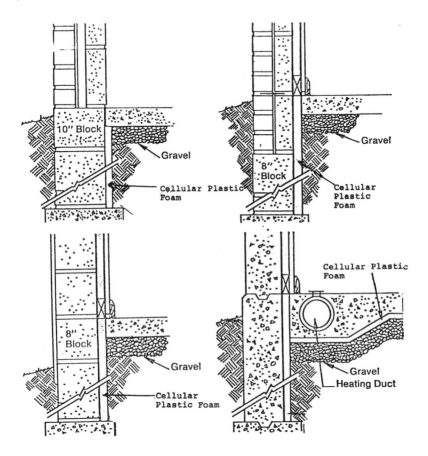

Source: BNL-50862

For the vertical wall portions of a foundation, the insulation is installed by adhering the board to the masonry block with a mastic or asphaltic adhesive and then backfilling. For floor slabs, the boards are placed on the prepared earth grade and the concrete is poured upon the insulation. This assembly is primarily used for new construction, but there are instances where rigid foam may be added to the outside of foundation walls on the area of the wall exposed above grade line. In this case, some type of barrier coating must cover the foam to prevent degradation from ultraviolet light. Below grade application of insulation around the perimeter edge would require some excavation which may be practical in some instances and impossible in others.

The foundation wall of a building can be sizable heat loss area. The new ASHRAE 90-75 standard includes requirements as to maximum U-values for floor slabs depending on the climate. ASHRAE also indicates that the insulation should continue down the vertical foundation wall for a minimum of two feet. In the residential building area, the FHA Minimum Property Standards have set requirements for perimeter heat loss from slab-type floors. To meet these standards, anywhere from ¾-inch to 2-inch thick rigid plastic foam is used depending on the winter design temperatures.

Few manufacturers present claims with respect to the percentages of heat loss saved by this application, but all show the U-values of various combinations which can be used to calculate heat loss.

Resistance to fire is not a consideration in this application due to the nature of the installation. When the plastic foam is used on the interior surface of the foundation wall in a crawl space, all manufacturers recommend using the same thermal barrier requirements as used elsewhere in the building.

The advantages of this type of insulation assembly are many. Not only do they retard heat loss from a highly conductive part of the building envelope, but their inclusion limits condensation formation on floors and improves the human comfort factor in a building. The only limitation of the system is the difficulty presented for retrofit in many instances.

SURVEY OF COMMERCIAL/INDUSTRIAL ASSEMBLIES

A survey of 30 architects located in all sections of the country was made to determine their requirements and use of thermal insulation in nine geographic areas of the United States. The cross section of contacts made were as follows: Pacific, 4; Mountain, 3; West South Central, 4; West North Central, 2; East North Central, 5; East South Central, 2; South Atlantic, 2; Middle Atlantic, 6; and New England, 2. 75% of the architects interviewed were concerned with the complete envelope, and in several cases took into consideration insulation of the floor slab as well. In addition to the envelope, other facets of this survey were:

(1) Type of structure − Steel, 90%; wood, 5%; concrete, 40%. The overlap indicates some firms built both steel and concrete structures or a combination of both.
(2) High rise or low rise − Low rise structures comprised 75% of the total, while the remaining 25% were high rise.
(3) Building type and/or use − Commercial buildings such as shopping centers, schools, churches, and office buildings comprise 75% of building types. Residential, such as apartments, barracks, nursing homes, were 10% and the remaining 15% were industrial type such as small manufacturing plants.
(4) New construction or retrofit − New construction amounted to 80% of the work done by these architects. The 20% of retrofitting was primarily due to reroofing. In almost every instance, additional insulation was added when a new roof was installed. It was noted that little or no tear-off of old roofs was done, as protection of

the building interior was critical, and any tear-off could expose this area to water damage. In some instances of retrofitting, the need for added insulation was critical and insulation in the form of batts or rolls was installed over a suspended ceiling. Occasionally, a cellulosic spray was used to increase thermal value.

(5) Type of insulation — For roof/ceiling construction, the type of insulation primarily used was in rigid board form. This was installed mostly over steel roof decks, 75%; wood, 15%; and concrete decks, either plain or insulating, 10%. The board types consisted of perlite, mineral fiber, polystyrene, urethane, foam glass, and combination perlite and urethane. Spray urethane was a very minor factor, but other reports indicate increasing use in thermal refitting. Insulating, leveling fills such as asphalt and perlite, perlite concrete, vermiculite concrete, or insulating concrete made up 20% of insulation types.

(6) Wall construction — Those concerned with the complete building envelope, about 75%, employed masonry systems for sidewalls. The most popular construction was block walls with an insulated air space between exterior wall and interior finish. The insulation usually was a rigid foam, either urethane or polystyrene. Some designs called for batt type insulation between furring studs to which was fastened the interior finish. One architect reported he used hollow block walls into which were poured vermiculite pellets.

(7) Design criteria for insulation (ASHRAE 90-75, local codes, or manufacturers' recommendations) — All architects were familiar with ASHRAE 90-75, but felt the standards were too restrictive. In some instances, the architect set his own standard or the customer for whom he designated dictated what overall U-value was to be achieved. If local codes had adopted U-values, such as Title 24 in California, then the architects were required to meet these codes. If not, then a typical design criteria was a U of 0.08 for walls and 0.10 for ceilings. Most architects accepted manufacturers' published values for insulation; however, some had tests conducted to insure that the thermal properties as published were correct.

(8) Design criteria for fire protection (Factory Mutual, Underwriters', burning and toxicity of foams) — As steel decks were used by 75% of those surveyed, Factory Mutual approved constructions played a major role in determining what design parameters were set. Factory Mutual seemed to be much more of a force than Underwriters'. If the constructions were Factory Mutual approved for fire from the underside, there was little or no particular concern for the roof covering having to meet any specific Underwriters' Laboratories classification. It was apparent, however, that designs were developed for roofing systems rated at no less than Class B for the type building under discussion. Combustion and toxicity of foam insulations did not seem to be a major concern since foams were encased in a roofing system or in a wall cavity.

THE USE OF WATER VAPOR BARRIERS WITH BUILDING INSULATION

The normal heating and cooling of the interior spaces of buildings results in a significant potential for the undesirable transmission and accumulation of moisture within building envelope systems. Water vapor barriers are materials which are used in buildings to provide a resistance to the transmission of water vapor in the building envelope under specified conditions.

Moisture transfer in building envelope systems is caused by differences between the indoor and outdoor absolute humidity. The rate of moisture transmission through building envelopes is proportional to the overall permeability of the envelope structure. Water vapor passing through a building envelope structure will condense in the structure at any point where the local temperature is below that of the local dew point temperature. The presence of condensed water in insulation will markedly affect its thermal resistance and the presence of condensed water for extended periods of time can result in the physical degradation of many common building materials.

Insulation in heated buildings reduces the heat loss from the building and lowers the temperature of the outer element of the building structure. This increases the likelihood of condensation if no measures are taken to prevent the movement of water vapor from the interior of the building. Similarly, buildings with air conditioning may experience condensation in the inner elements of the building structure due to the propagation of moisture from the environment into the building envelope.

The increased use of insulation in buildings, along with other measures such as improved weather-stripping have significantly reduced air infiltration-exfiltration rates and, thus, have increased the importance of an effective vapor barrier. Reducing the rate of air replacement and thus reducing the rate of water vapor removal from the building results in increased humidities in the conditioned spaces.

Types of vapor barriers utilized with building insulation systems may be classified as structural, membrane, or coatings. Structural water vapor barriers would include rigid insulations which are relatively impervious to water. Membrane barriers include metal foils, laminated foil and treated papers, coated felts and papers, and plastic films or sheets. These membrane barriers are flexible and are supplied in roll form or as an integral part of an insulation. Coating barriers are found in a variety of semifluid or fluid forms and may be asphaltic, resinous, or polymeric in composition.

The American Society of Heating, Refrigerating, and Air Conditioning Engineers, in recognition of the distressing results its omission may bring, has tentatively recommended that the walls of every well-constructed modern dwelling include a vapor barrier when the construction includes any material which would be damaged by moisture or freezing. In applying a vapor barrier to a wall, ASHRAE recommends that certain fundamental principles should be followed. They are that the vapor barrier should be placed as near to the warm surface of the wall as practicable and that it should be continuous with no direct openings through the barrier. Field data which indicate that a vapor barrier with only a small break can result in excessive moisture levels substantiate this recommendation.

Mineral fiber insulation materials are normally installed in new wooden frame buildings in a batt form with an integral vapor barrier. The types of barriers used include compositions of kraft paper, aluminum, and asphalt. The ASHRAE Handbook of Fundamentals notes that batt or blanket insulation with an integral vapor barrier is reasonably adequate in ceilings if installed by fastening the batt to the joists by nailing flanges which are provided. However, it is further noted that in the general installation of this type of insulation—vapor barrier system, the edges should be lapped over framing members and stapling to the sides of framing should be avoided. If stapling to the sides of the framing is unavoidable, a separate vapor barrier, such as polyethylene plastic sheeting, is recommended for best results.

Applications of loose-fill mineral fiber or cellulose insulation in either new construction or existing building ceilings requires the installation of a vapor barrier. Polyethylene sheeting is commonly used for this purpose.

Plastic foam sheathing applied to the exterior of framed buildings in conjunction with a cavity fill insulation requires a good vapor barrier on the interior wall since the foam sheathing would tend to prevent normal drying out of the wall during favorable weather conditions. Experience with systems composed of plastic foam sheathing, mineral fiber batts and polyethylene, kraft-asphalt, or foil vapor barrier indicates moisture control is adequate.

Retrofit of walls with insulation poses a problem with respect to constructing an adequate moisture barrier. Studies by the Forest Service of the U.S. Department of Agriculture (2)(3) have concluded that the addition of insulation to walls of an older house with no vapor barrier subject to the climate of Madison, Wisconsin, may not cause moisture problems.

However, it is noted that this is a marginal situation and will vary with tightness of the house and the habits of the occupants. Further, their study concludes that where no visible problems occur, moisture decreases the thermal efficiency of the insulation and the use of mechanical humidification presents special problems. The wall moisture data acquired by the Forest Service nevertheless does demonstrate the effectiveness of aluminum paint applied over the plaster even where mechanical humidification is used.

DYNAMIC INSULATION

Standard methods of thermal insulation are based on the principle of preventing the movement of air in a confined space. A new method of building insulation, which has been developed with the support of the National Swedish Board for Technical Development depends on moving rather than stationary air (4). The new system of dynamic insulation depends on the use of a controlled slow flow of air inward through the wall insulation opposite to the outward flow of heat. The incoming air absorbs the escaping heat and returns it to the interior. Originators of this concept claim that in houses with dynamic insulation it is possible to regulate the outward flow of heat so that the effective R-value of the walls is very large.

The proper function of this system requires a large enough air flow rate through the insulation to minimize conduction to the outside. However, since the air drawn into the building will necessarily be exhausted in some manner, there is the need to minimize this air flow to minimize heat withdrawn from the building. It is claimed that at an air flow of 6 ft/hr, the apparent thermal conductance of the wall diminishes to near zero and results in an air change rate in a typical residence of approximately half the volume per hour. Achieving the proper air flow into the building requires an unconventionally tight house of special construction. Any air drawn into the house at any point other than through the insulation will result in degraded overall thermal performance of the structure. During the summer months the same dynamic heat recovery principle applies. Heat flow into the building could be reduced by reversing the direction of air flow. Additionally, solar gain on the outside wall of the structure during the winter would possibly benefit the overall heat balance on the building when utilizing this system.

The Swedish originators are continuing development of this concept.

REFERENCES

(1) *Sweets Catalog Service, Light Construction Catalog File,* McGraw-Hill, Sweets Division, New york, 1977.
(2) Duff, J.E., *Forest Prod. J.,* 18, 60, 1968.
(3) Sherwood, G.E. and Peters, C.C., "Moisture Conditions in Walls and Ceilings of a Simulated Older Home During Winter," USDA Forest Service Research Paper, FPL 290, 1977.
(4) Sundell, J., "Swedish Experience in Energy Efficient Housing," Energy Efficiency in Wood Building Construction, Forest Products Research Society, Chicago, IL, November 8-10, 1977.

RETROFITTING RESIDENTIAL BUILDINGS

The material in this chapter was excerpted from reports
prepared by Brookhaven National Laboratory with the
assistance of Dynatech R/D Company (BNL-50862) and
the Division of Buildings and Community Systems, Office
of Conservation and Solar Applications of the Department
of Energy (DOE/CS-0051).

IMPORTANCE OF RETROFITTING

In 1973, at the start of the energy crisis, it was estimated that more than 40 million
single-family residences could be retrofitted by the addition of insulation in ceilings, walls, and floors, storm windows and doors, caulking and weatherstripping
and similar measures to reduce the amount of energy used for space heating and
cooling and water heating by nearly 20%. A study conducted by the Office of
Businesses Research and Analysis of the U.S. Department of Commerce in 1977
estimated that 25.5 million housing units were in need of retrofitting; an additional
8 to 9 million units were voluntarily retrofitted since 1973-74.

The application of weatherization materials to an existing residential building is
not totally without risk of hazard to the occupants of the building or to the
structure itself. However, these risks can be minimized through the use of quality materials designed to respond to the rigors of the environment in which they
are placed.

The Division of Buildings and Community Systems (BCS), Office of Conservation
and Solar Application of the Department of Energy, cognizant of potential risks
associated with the weatherization of existing buildings and the high level of activity presently being undertaken in the marketplace, drafted material criteria to
assist manufacturers, installers, and consumers in selecting materials for retrofitting
residential buildings in a manner which will guard, to the maximum extent possible, the safety of the occupants, and the effectiveness of the measures installed.
The criteria for the following list of materials are described in the next section.

Mineral fiber blanket or batt insulation
Mineral fiber loose-fill insulation
Cellulose loose-fill insulation
Perlite thermal insulation
Vermiculite thermal insulation
Polystyrene insulation board
Polyurethane and polyisocyanurate insulation board
Aluminum foil reflective insulation

BCS recognizes that good installation practices are also essential to ensure the safety and effectiveness of all energy conservation measures. Consequently, BCS has drafted installation practices that were prepared to meet the following considerations:

That the installed materials serve the function they were designed to serve, i.e., conserve energy

That the installed materials do not increase the potential of a hazardous situation within a residence. Hazardous situations include both fire and structural damage.

That consumers are offered protection against fraud.

That installation is conducted in such a way so as to assure the durability of the material.

RECOMMENDED CRITERIA FOR INSULATION MATERIALS

The materials and products considered appropriate for the retrofit of homes include: thermal insulation, storm windows and doors, caulks and sealants, weatherstripping, water heater insulation, clock thermostats, and multiglazing when used for retrofitting residences to save energy. This section gives criteria by which the appropriateness of insulating materials could be determined. It has been shown that proper use of all these materials and products is effective in conserving energy and generally cost effective. These criteria are based on factors such as thermal performance, fire safety, health safety, structural integrity, durability, quality, use and ease of installation.

Thermal insulation is a material or assembly of materials used primarily to provide resistance to heat flow. The materials listed previously are those considered suitable for use in retrofitting residences to provide increased thermal resistance.

They should meet requirements for heat flow resistance (i.e., thermal resistance), fire safety, noncorrosiveness, and resistance to moisture absorption, odor emission and fungi. Insulation materials should be considered acceptable for retrofitting residences only if they conform to the applicable Federal Specification and also conform to the fire safety requirements described below. Federal Specifications are available which contain requirements for most of these material properties.

Federal Specifications do not, in many cases, contain adequate fire safety requirements. Fire safety requirements involve consideration of burning characteristics of the insulation such as its ignitability, rate of heat release, smoldering, surface flame spread, and flame resistance permanency. Installation instructions should be included by manufacturers with thermal insulation materials to facilitate their application in a safe and effective manner.

Urea-Formaldehyde Foam Insulation

Following the advice of the interagency task force on material standards, BCS has considered developing a material criteria for urea-formaldehyde (UF) foam insulation. There are two problems, however, which must be resolved before such a criteria can be developed:

> (1) UF foam shrinks during setting, reducing the insulation's thermal resistance. The shrinkage rate and degree of thermal resistance reduction have not yet been adequately measured.

> (2) Formaldehyde vapor is sometimes given off from UF foam. This vapor has a noxious odor and may pose a health hazard to occupants of house insulated with UF foam.

No acceptable level of the maximum allowable concentration of formaldehyde vapor in UF foam-insulated houses has yet been set. The ASTM task group on UF foam has not yet set an acceptable level for formaldehyde vapor concentration. Because there is a need for better data on the shrinkage and formaldehyde vapor problems associated with UF foam application, BCS has initiated several studies to gain a better understanding of these properties.

BCS is also assembling information from various federal and state agencies on UF foam and the associated problems. The Consumer Product Safety Commission (CPSC) held public hearings on UF foam odor complaints in Denver, Colorado. Evidence was presented that the incidence of reported problems in all installations has been less than 0.1%. The Massachusetts Department of Consumer Affairs held a meeting on UF foam in Boston chaired by DOE. Representatives attended from DOE, HUD, NBS, Federal Trade Commission, Center for Disease Control, EPA, and the CPSC and from the states of Wisconsin, Massachusetts, Connecticut, New Jersey, New Hampshire, and Maine. BCS is continuing to solicit information on the formaldehyde vapor and shrinkage problems associated with UF foam applications.

Fire Safety Requirement for Thermal Insulation Materials

Some Federal Specifications currently require that the fire performance of the various insulation materials be evaluated according to the test procedures described in ASTM E84 "Surface Burning Characteristics of Building Materials." The National Bureau of Standards does not consider this procedure to be appropriate for assessing the response of insulation to flame and heat in every application.

Based on this NBS determination, the BCS recommends that certain recently developed test procedures be required in addition to the existing one (E84). These fire safety testing requirements differ for various insulation materials. The recommended requirements also differ according to the area of application, that is, whether the insulation is installed in an open attic floor, an enclosed space such as a wall or floor cavity, or a wall or ceiling where the insulation will be exposed. Table 6.1 contains the BCS recommended fire safety requirements for insulation.

Table 6.1: Recommended Fire Safety Requirements of Insulating Materials

Insulation	. . . . Area of Application		
	Open Attic Floors	Enclosed Spaces	Exposed Insulation in Walls or Ceilings
Mineral fiber blankets/batts with membrane covering	CRF, SC	SC	CRF, SC, FS ≤150 SD ≤50
without membrane covering	CRF, SC	SC	CRF, SC
Mineral fiber loose-fill	CRF, SC	CRF, SC	NR
Cellulose loose fill	FS ≤25, FRP, SD ≤50	FS ≤25, FRP, SD ≤50	NR
Perlite	—	—	—
Vermiculite	—	—	—
Polystyrene board	NR	FS ≤75, SC	NR
Polyurethane and poly- isocyanurate board	NR	FS ≤75, SC	NR
Aluminum foil reflective insulation	CRF, SC	SC	FS ≤150

Note:
CRF = Critical radiant flux
SC = Smoldering combustion test
FS = Flame spread classification

SD = Smoke developed rating
NR = Not recommended for application
FRP = flame resistance permanency test

Source: DOE/CS-0051

Critical Radiant Flux: Critical radiant flux is the level of incident radiant heat energy on the attic floor insulation system corresponding to the furthest point at which flame propagation ceases. For insulations, the critical radiant flux should be equal to or greater than 0.12 W/cm^2.

The critical radiant flux should be determined by the attic floor radiant panel test procedure described in Federal Specification HH-I-515D. This method of test provides a basis for evaluating the surface flame spread behavior of an attic floor insulation in a building attic. The fundamental assumption inherent in the test is that "critical radiant flux" is one measure of the sensitivity to flame spread of installed attic floor insulation.

Smoldering Combustion: This test determines the tendency of the insulation to support and propagate smoldering combustion subsequent to exposure to a standard localized ignition source. When tested for smoldering combustion, the insulation should meet the following requirements:

> Weight loss should not exceed 15% of the initial weight
> No evidence of flaming combustion shall be observed

The smoldering combustion test should be conducted according to the procedure described in Federal Specification HH-I-515D.

Surface Burning Characteristics and Smoke Developed: The ASTM E84-77 flame spread test method should be the basis for evaluating surface burning character-istics and smoke developed. This test method is recommended for polystyrene, polyurethane, and polyisocyanurate insulation board.

It is also recommended for aluminum foil insulation and mineral fiber blanket insulation having membrane covering if these materials are exposed after installation in walls or ceilings.

Assessment: Although a degree of material combustibility is allowed, the intent of these fire safety requirements is to allow insulating materials which are not more combustible (or flammable) than acceptable existing construction and insulation materials, and to preclude any increased fire hazard due to the retention of heat from energy-dissipating objects.

In areas where occupants are likely to be engaged in normal activities, the insulation should perform its intended function without the increased risk of ignition, rapid flame spread, and heat and smoke generation. Insulation in concealed spaces may be a particular fire problem due to its inaccessibility for fire fighting.

Although the primary fire safety properties of insulation relate to ignitability and rate of heat release, standard test methods for these properties do not exist. The critical radiant flux and smoldering combustion test method should be used to judge the fire safety of mineral fiber blankets, mineral fiber loose-fill, cellulosic loose-fill, and aluminum foil reflective insulation. Federal Specification HH-I-515D has already incorporated such tests. Federal Specifications, HH-I-521E and HH-I-1030A are undergoing revision, and it is anticipated that the critical radiant flux and smoldering combustion tests will be incorporated in the revisions.

Many insulation materials, including those consisting of cellulose, plastic foam and fibrous glass (containing organic binder) are combustible materials which will burn and release heat, smoke and gases, especially when exposed to continuous large fire sources. Additional information on the rate of heat release and on performance of these insulation materials in full-scale room fire tests should be developed.

INSTALLATION PRACTICES

Installation practices are proposed which apply to the following generic classes of insulation determined suitable for residential retrofit use:

> Organic (cellulosic or wood fiber) and mineral fiber loose-fill thermal insulation
>
> Mineral fiber batts and blankets thermal insulation
>
> Organic cellular rigid board thermal insulation
>
> Mineral cellular loose-fill thermal insulation, and
>
> Reflective insulation

General Issues Relating to Insulation

The installation practices presented herein are based on the best available technical information. However, several issues existed which required general determinations to be made. These determinations, as well as the technical support on which they are based are presented below:

(a) The installation practices discourage the application of insulation material in wall cavities which have been previously insulated. This is based upon four criteria:

(1) Additional wall insulation is unlikely to be cost-effective. To a large extent, the cost of additional insulation is attributable to the labor involved in the installation process. The addition of such material could not provide energy savings sufficient to off-set the installation cost. This determination with respect to cost effectiveness is supported by an economic analysis performed by the National Bureau of Standards ("Retrofitting Existing Housing for Energy Conservation," Building Science Series 64, NBS, Department of Commerce, Washington, D.C., December 1974) which concluded "...if any insulation exists (in walls), blown-in insulation is not practical and unlikely to be cost-effective."(Ibid. p. 45).

(2) Allowing additional wall insulation increases opportunities for consumer fraud since many homeowners do not know whether or not they have existing wall insulation.

(3) Installing wall insulation in a cavity where some exists greatly reduces the potential for an effective application since existing insulation restricts the access to the wall cavity.

(4) Additional insulation material may not be physically or chemically compatible with insulation in place. When a wet material is installed, for instance, it may saturate dry materials rendering them ineffective.

(b) BCS is concerned about the effects of thermal insulation placed both over and under electrical wiring. Laboratory tests conducted by the National Bureau of Standards (NBS) show that when thermal insulation is placed around electrical wiring, heat dissipation is restricted and resulting wiring temperatures greatly exceed those allowed by the National Electrical Code (NEC) or Underwriter's Laboratory (UL) (See NBSIR 78-1477, "Exploratory Study of Temperatures Produced by Self-Heating of Residential Branch Circuit Wiring When Surrounded by Thermal Insulation"). Both NEC and UL require that the type of electrical wiring often found in existing residences not be operated at temperatures in excess of 140°F.

When parallel nonmetallic sheathed cables carrying 135% of rated current were placed between two layers of R-11 insulation, NBS found that temperatures were more than double the allowed limit. The tests were considered conservative since they were conducted with new wires over relatively short runs. In addition, current rated at 135% may not be as common as 150% in those situations where over-fusing occurs. Overheated wiring circuits may ignite electrical insulation or adjacent building materials, or they may lead to gradual deterioration of the electrical insulation which could result in ignition from arcing or short-circuits. The severity of the situation requires BCS to take precautionary measures. Because there is sufficient evidence to indicate to BCS that a problem might exist, BCS is considering the inclusion in the installation practices of a statement similar to the one that follows:

"Attics with exposed wiring should be treated in one or more of the following ways:

  (1) Insulation should be placed in a ceiling to a uniform depth up to the underside of electrical wiring, provided that a reasonably unstructured open air space is permanently retained above such wires; or

  (2) Insulation in the ceiling should be placed only between those joists which have no wiring extending perpendicular to or parallel in the joist area; or

  (3) Insulation in the ceiling should have barriers installed below and around the sides of the wires in excess of the full height of the insulation. The barrier should form an air space which will permanently restrain the thermal insulation from covering the electrical conductor(s). The barrier should be permanently left in place as a retaining element and made of such material that is consistent with local building requirements for durability, safety, fire, moisture resistance, rodent and fungus protection.

Provisions (1) through (3) noted above do not apply to wiring which is attached parallel along the joist."

This provision will ensure that wire temperatures under prolonged normal rated loading conditions do not exceed the values specified in the NEC for the appropriate type of wiring. In the meantime, BCS, and NBS are jointly conducting a survey which will examine wiring temperatures and degradation of wiring in actual homes. If this survey shows that wiring in actual homes does not, for some reason, reach and sustain the temperatures encountered in the laboratory environment, a provision similar to that noted above, will not appear.

(c) BCS is concerned about how insulation, with and without vapor barriers, affects moisture transmission and accumulation. Many experts feel that installing insulation without a vapor barrier may substantially increase moisture accumulation in wood frame walls. If moisture accumulation is increased, it is unclear how serious resulting conditions will be. It is clear, however, that a large number of factors affect moisture transmission including: geographic location, permeance of interior and exterior wall parts, infiltration through the wall, indoor humidity level, etc. There is no agreement, however, as to the degree to which each of these factors influence moisture transmission.

BCS is continuing to: (1) evaluate the extent of moisture problems, (2) better define the relationship between insulation and moisture transmission, (3) determine potential long-term deterioration resulting from moisture accumulation, and (4) determine how to better control moisture transmission through the indoor environment.

(d) The insulation installation practices recommend minimum ventilation of attic spaces before insulation can be installed. The condensation of moisture may occur as a result of cooler outside air meeting warmer attic air. This moisture settles in the roof, wood beams or insulation causing damage to the structure or diminishing the thermal effectiveness of the insulation; it is therefore necessary to provide for sufficient air circulation to evaporate the moisture and prevent condensation from accumulating.

(e) The insulation installation practices recommend the installation of vapor barriers in attics where no insulation is in place as well as in attics where existing insulation is to be replaced. In addition, where insulation does exist or is to be installed in wall cavities, this section states that a vapor barrier should be applied in Zone 1 as depicted in Figure 6.1. In Zones 2 and 3 of Figure 6.1, the application of a vapor barrier is recommended. Since a large amount of vapor results from the residential use of water, this requirement would reduce the transmission of this vapor to insulated spaces where it might otherwise saturate the insulation material causing a reduction in its thermal effectiveness or causing structural damage in adjacent areas.

Because so many questions exist regarding the effects of moisture transmission and accumulation on both insulation and structure BCS is considering limiting suggested areas for vapor barriers to only kitchens, bathrooms, and laundry areas in conditioned spaces in Zone 1.

(f) The installation practices recommend that blocking be installed around all heat-producing appliances when loose-fill insulation is installed to provide a 3" clearance. This provision is in compliance with the National Electric Code. In the case of recessed light fixtures, there should be no insulation material above such fixtures, except where the fixture is approved for mounting in contact with insulation by the manufacturer and the device is so labeled. The heat build-up which might occur as a result of closing off the ventilation accesses provided on recessed lighting fixtures and other heat-producing devices may be sufficient to cause smoldering which can spread to ignitable materials within the fixture and adjacent structural members.

In addition, where insulation is placed above recessed lighting fixtures, sufficient heat can build up to cause ignition of some insulation materials including vapor barriers which are provided as an integral part of the material or installed independently of the material. The requirement would also apply to installing insulation materials in order to avoid contact with other heat sources such as flues to ensure a heat build-up does not occur in these locations causing ignition of adjacent materials.

Special Determinations

The installation practices do not include installation of dry cellulose which is mixed and sprayed with water as it is applied or cellulose which is installed as a premixed wet material. The omission is predicated upon a determination that this type of installation may result in:

> Increased corrosion of other materials within the cavities insulated;

> Excessive accumulations of water in blind wall cavities because of the difficulty in judging the proper mix of water and insulation material, thus resulting in the possibility of structural damage; or

Figure 6.1: Condensation Zones in the United States

Source: DOE/CS-0051

Excessive accumulations of water in cavities because of the
presence of an exterior vapor barrier which has been applied,
i.e., oil-based paint or similar low permeance coatings, which
prevents the material from drying, thus diminishing its thermal
effectiveness and possibly causing other problems such as
paint peeling and structural decay.

Because of the lack of specific information concerning the actual potential for
structural damage or loss of thermal effectiveness as a result of wet installation
processes, BCS will further investigate the results of wet installation through its
research facilities. After a review of the findings which result from such research,
the Department will reassess its position with respect to the prohibition against
the wet process installation of cellulose.

PROCEDURES PRIOR TO AND FOR INSTALLATION

Organic (Cellulosic or Wood Fiber) and Mineral Fiber Loose-Fill Insulation

This section describes standard practices for the installation of organic and min-

eral fiber loose-fill insulation in ceilings, attics, and wall cavities of existing
residential buildings. It applies only to the installation of dry loose-fill
thermal insulation consisting of organic (cellulosic or wood fiber) materials or
mineral fiber by pneumatic or manual means. It does not apply to material
installed in a wet condition or where liquid is added at any stage of the installation process.

Walls: Where there is existing insulation in wall cavities, no additional insulation
should be installed. Buildings located in Zone 1 of Figure 6.1 should have a
vapor barrier, such as paint and wallcoverings which are labeled by the manufacturer as having a perm rating of one or less and are applied in strict accordance to the manufacturer's instructions, on the winter-warm side of exterior
walls to be insulated. It is recommended that buildings in Zones 2 and 3 of
Figure 6.1 have a similar vapor barrier on the winter-warm side of exterior walls
to be insulated.

Attics and Ceilings: Ventilation openings in attic areas to be insulated should
conform to one of the following requirements:

  (a) 1 ft^2 minimum of free ventilation area per 150 ft^2
       of attic space, if no vapor barrier exists in the attic;

  (b) 1 ft^2 minimum of free ventilation area per 300 ft^2
       of attic space if a vapor barrier does exist;

  (c) 1 ft^2 minimum of free ventilation area per 300 ft^2
       of attic space if at least 50% of the required ventilating
       area is provided with fixed ventilation located in the
       upper portion of the space to be ventilated (at least 3
       feet above eave or soffit vents) with the remainder of
       the required ventilation provided by eave or soffit vents.

For buildings located in Zone 1 of Figure 6.1 if there is no existing insulation
or if existing insulation is to be removed, a vapor barrier membrane should be
applied on the upper surface of the ceiling material. It is recommended that a
vapor barrier be applied in buildings in Zones 2 and 3 of Figure 6.1. For buildings in Zone 1, if there is existing attic insulation and no existing vapor barrier,
a vapor barrier such as paint and wallcoverings which are labeled by the manufacturer as having a perm rating of one or less and are applied in strict accordance to the manufacturer's instructions should be provided on the interior surface of the ceiling. It is recommended that a similar vapor barrier be provided
in buildings in Zones 2 and 3. In no case should a vapor barrier be applied on
top of existing insulation.

Rigid blocking should be installed in attics to restrain the loose-fill insulation
from falling into attic doors or other accesses. When the attic has soffit vents
at the eaves, rigid blocking should be installed to restrain loose-fill insulation from
clogging the vents thus restricting attic ventilation. Blocking should be installed
in such a way as to ensure free movement of air through soffit vents into the
attic. All vent openings in the attic should be covered with temporary blockings
prior to the installation of insulation to ensure no insulation material falls into
the vents.

General: The installed insulation should not be in contact with the ground or other sources of water and should only be installed between conditioned and unconditioned spaces for the purpose of energy conservation. Entry holes for pneumatic equipment should be located so as to permit the complete filling of wall cavities. A complete fill generally requires two openings per floor per stud space. More may be needed where obstructions are present within the cavity. Entry holes should be opened cleanly with a technique that permits refinishing with little or no change in the appearance or structural integrity of the wall. After the entry holes have been opened they should be used to check the wall cavity for fire stops and other obstructions which may require additional entry holes to assure complete filling of the cavity.

The pneumatic installation in ceiling areas should use the least amount of air possible to convey the insulation so as not to disturb adjacent insulation or create "pockets" that affect density. The attic side of access doors or panels should be fitted with insulation batt (or equivalent material) except where there is a retractable ladder or other equipment attached.

Mineral Fiber Batts and Blankets Thermal Insulation

This section describes standard practices for the installation of mineral fiber batt and blanket insulation in ceilings, attics, floors, and wall cavities of existing residential buildings.

Walls: Buildings located in Zones 1, 2 and 3 of Figure 6.1 should have a vapor barrier on the winter-warm side of exterior walls to be insulated.

Attics and Ceilings: Ventilation openings in attic areas to be insulated should conform to one of the following requirements:

    (a) 1 ft^2 minimum of free ventilation area per 150 ft^2 of attic space, if no vapor barrier exists in the attic;

    (b) 1 ft^2 minimum of free ventilation area per 300 ft^2 of attic space if a vapor barrier does exist;

    (c) 1 ft^2 minimum of free ventilation area per 300 ft^2 of attic space if at least 50% of the required ventilating area is provided with fixed ventilation located in the upper portion of the space to be ventilated (at least 3 feet above eave or soffit vents) with the remainder of the required ventilation provided by eave or soffit vents.

For buildings located in Zone 1 of Figure 6.1, if there is no existing insulation or if existing insulation is to be removed, a vapor barrier should be provided on the upper surface of the ceiling material. It is recommended that a vapor barrier be applied in buildings in Zones 2 and 3 of Figure 6.1. For buildings in Zone 1, if there is existing attic insulation and no existing vapor barrier, a vapor barrier such as paint and wallcoverings which are labeled by the manufacturer as having a perm rating of one or less and are applied in strict accordance to the manufacturer's instructions should be provided on the interior surface of the ceiling.

It is recommended that a similar vapor barrier be provided in buildings in Zones 2 and 3. In no case should a vapor barrier be applied on top of existing insulation.

Floors: Where insulation is to be installed beneath floors over crawl spaces, the ground surface should be covered with a vapor barrier. Crawl spaces should have a free ventilating area of one square foot for every 1,500 square feet of the ground area before insulation is applied. These spaces should be cross ventilated if possible. A vapor barrier should be provided on the winter-warm side of floors to be insulated in buildings located in Zones 1, 2 and 3 of Figure 6.1.

General: The installed insulation should not be in contact with the ground or other sources of water and should only be installed between conditioned and unconditioned spaces.

It should fit tightly between joists and studs on all sides. Insulation that is too long for a space should be cut to the correct size. If insulation is too short for a space, a piece should be cut to fill the void. Insulation should fit properly and shall not be doubled over or unnecessarily compressed. All joints and tears in the vapor barrier should be taped and sealed.

A 3-inch minimum air space should be maintained around motors, fans, blowers, heaters, and other heat-producing devices. This also applies to flues and chimneys when a backing is attached to the insulation.

A 3-inch minimum air space should be maintained around recessed lighting fixtures, ballasts, and wiring compartments. Insulation should not be installed above lighting fixtures or other heat-producing devices so as to cause over-heating of these devices. All devices which may require periodic servicing should remain accessible after the insulation is installed.

Walls: Insulation should be correctly sized for the cavity and fit tightly at the sides and both ends. Batts shorter than the cavity length should be tightly butt-jointed. When a vapor barrier is provided with the insulation it should be secured so as to avoid gaps and fishmouths.

Where a vapor barrier is not attached to the insulation to be installed, a separate vapor barrier, facing the winter-warm side, should be applied. The interior side of the insulation should be covered with a suitable finish material so that no portion of the insulation is exposed to habitable spaces.

Attics, Ceilings, and Floors: If there is a vapor barrier it should be placed on the winter-warm side.

When installing insulation around bridging or cross bracing of ceiling or floor joists care should be taken to assure that the insulation material is fitted tightly around these obstructions and that there are no gaps in the insulation.

Insulation installed between floor joists should be held in place with either wire fasteners, galvanized wire or nylon mesh or galvanized screen held in place by stapling or nailing, galvanized wire lacing held in place by stapling or nailing, or, if the insulation is provided with a reverse-flange, stapling to the joists.

Soffit vents should not be covered with insulation nor in any way should attic ventilation be restricted.

Insulation should be installed around vents which open into the attic to ensure free movement of air through the vent into the attic.

The attic side of access doors or panels should be fitted with insulation batt (or equivalent material) except where there is a retractable ladder or other equipment attached.

Organic Cellular Rigid Board Thermal Insulation

This section describes standard practices for the installation of organic cellular rigid board thermal insulation on flat roofs, concrete slab floors, slab-on-grade or foundation perimeters, heated basement or other masonry walls, and as exterior sheathing for existing residential buildings. It does not apply to types of rigid board intended for heavy duty or nonresidential uses, such as Polystyrene Type III, Class B (compressive strength of 100 psi).

Walls: When installing rigid board on the interior side of exterior masonry walls, fire stops should be placed at 8 to 10 foot intervals in horizontal and vertical directions and at floor level and ceiling level. When used as exterior sheathing on buildings of more than one story, the insulation board should be fire stopped at each floor level and at the ceiling of the uppermost story and in between dwelling units.

If the exterior perimeter and subgrade walls are to be insulated, a two foot (minimum) trench should be dug adjacent to the wall before installation. The material selected should be suitable for underground and outside applications.

Roofs: It should be determined whether roof areas to be insulated have sufficient load-bearing capacity to support the insulation board installation and its covering materials. The condition of the existing waterproof membranes and flashing should be inspected and any required repairs performed.

Installation Procedures—General: The installed insulation should not be in contact with the ground or other sources of water, except in the case of slab perimeter installations where insulation board specifically designed for this application is used. Insulation board should only be installed between conditioned and unconditioned spaces; it should not be installed on the interior walls of flue-like structures such as stairwells.

Insulation board materials should not be exposed to sunlight for longer periods than recommended by the manufacturer. Installed insulation board should be continuous and without gaps. If a vapor barrier is recommended and is provided as an integral part of the insulation board material, the joint of two insulation boards should be sealed to assure continuous moisture protection.

Basement and Masonry Walls: Insulation board installed on the interior of basement and masonry walls should be covered with a minimum of ½" gypsum wallboard or other suitable material having a fire rating of 15 minutes minimum (per ASTM-E-119-73, "Standard Methods of Fire Tests of Building Construction and Materials"). The boards should be secured to the walls firmly either mechanically or with adhesives in accordance with the manufacturer's recommendations.

Exterior Sheathing: Insulation board materials which are acceptable for use as exterior sheathing may be applied to the existing envelope of a building and then covered with new siding. However, such insulation and new siding added to the existing wall should be firmly supported on the structure frame or wall. Existing wood studs may be considered structure frame and existing wood siding or masonry wall may be accepted as structure wall. If a vapor barrier is provided as an integral part of the insulation board material, a separate vapor barrier is recommended on the winter-warm side of the wall. Adequate ventilation as recommended by the insulation board manufacturer may be provided in lieu of a separate vapor barrier. Insulation materials used on exterior wall surfaces should be suitable for outdoor application.

Slab-on-Grade and Foundation Perimeters: The insulation board should be secured to the wall in accordance with the manufacturer's recommendations. All insulation board above the ground line should be covered with an exterior wall finish. Trenches should be backfilled and tamped to proper grade and consistency.

Concrete Floors: The insulation board should be installed as recommended by the manufacturer. A suitable subflooring material such as $\frac{1}{2}$" exterior grade plywood should be secured on top of the insulation board in accordance with the insulation manufacturer's instructions before applying the final finishing floor (tiles, carpets, etc.).

Flat Roofs: A suitable roof covering material should be applied on top of the insulation board. If the loading permits, gravel and binder of specified size and weight per unit area as recommended by the insulation board manufacturer may be used in lieu of roof covering.

Mineral Cellular Loose-Fill Thermal Insulation

This section describes standard practices for the installation of perlite and vermiculite mineral cellular loose-fill thermal insulation in existing residential buildings. It applies only to the installation of mineral cellular loose-fill thermal insulation in attic floors and various masonry cavities (concrete block, blockbrick veneer, and masonry walls wherein the wall cavity opens into the attic) by means of pouring application. It does not apply to installation by pneumatic machinery processes.

Walls: Where there is existing insulation in wall cavities, no additional insulation should be installed. Buildings located in Zone 1 of Figure 6.1 should have a vapor barrier, such as paint and wallcoverings which are labeled by the manufacturer as having a perm rating of one or less and are applied in strict accordance to the manufacturer's instructions, on the winter-warm side of exterior walls to be insulated. Buildings in Zones 2 and 3 should have a similar vapor barrier on the winter-warm side of exterior walls to be insulated.

Attics and Ceilings: Ventilation openings in attic areas to be insulated should conform to one of the following requirements:

> (a) 1 ft² minimum of free ventilation area per 150 ft²
> of attic space, if no vapor barrier exists in the attic;
>
> (b) 1 ft² minimum of free ventilation area per 300 ft²
> of attic space if a vapor barrier does exist;

(c) 1 ft² minimum of free ventilation area per 300 ft²
of attic space if at least 50% of the required ventilation area is provided with fixed ventilation located
in the upper portion of the space to be ventilated
(at least 3 feet above eave or soffit vents) with the
remainder of the required ventilation provided by
eave or soffit vents.

For buildings located in Zone 1 of Figure 6.1, if there is no existing insulation
or if existing insulation is to be removed, a vapor barrier membrane should be
applied on the upper surface of the ceiling material. It is recommended that a
vapor barrier be applied in buildings in Zones 2 and 3. For buildings in Zone 1,
if there is existing attic insulation a vapor barrier, such as paint and wallcoverings which are labeled by the manufacturer as having a perm rating of one
or less and are applied in strict accordance to the manufacturer's instructions
should be provided on the interior surface of the ceiling. It is recommended
that a vapor barrier be provided in buildings in Zones 2 and 3. In no case
should a vapor barrier be applied on top of existing insulation.

Rigid blocking should be installed in attics to restrain the loose-fill insulation
from falling into attic doors or other accesses. The attic side of access doors
or panels should be fitted with an insulation batt or equivalent material except
where there is a retractable ladder or other equipment attached.

General: The installed insulation should not be in contact with the ground or
other sources of water. Insulation should only be installed between conditioned
and unconditioned spaces for the purpose of energy conservation.

When installing insulation above gypsum board ceilings or above existing insulation which is resting on a gypsum board ceiling the new insulation installed
should not be an amount greater than that which would cause the total weight
of the existing insulation and new insulation per square foot of ceiling area to
exceed the loads specified in Column 3 of Table 6.2.

Table 6.2: Maximum Allowable Loads

....(1).... Gypsum Board Ceiling Thickness, inch	.......(2)........ Frame Spacing, inches	...(3).... Total Allowable Load, psf
½	24 O.C.	1.3
½	16 O.C.	2.2
⅝	24 O.C.	2.2

Source: DOE/CS-0051

When insulating the cavity of an enclosed attic floor, the attic floor should be
removed to facilitate proper installation and coverage.

Reflective Insulation

This section describes requirements for the installation of reflective (aluminum
foil) insulation under floors of heated spaces over unheated crawl spaces and un-

heated basements in existing residential buildings.

Preinstallation Inspection and Preparation: In the areas where insulation is to be installed, all visible wiring, junction boxes, and other metallic or electrical equipment should be identified and examined. If the wiring is found to have frayed, cracked, deteriorated or missing electrical insulation, or if equipment is not grounded, reflective insulation should not be installed until such conditions have been corrected by personnel qualified and approved for such work.

The installer should ensure that all electrical equipment in the building is grounded. Reflective insulation (aluminum foil) should not be installed in existing buildings in which electrical equipment is not grounded.

The installer should identify air supply and return ducts, pipes, electrical wires and other obstructions located in spaces between floor joists, over crawl spaces and unheated basements for special installation considerations. All metallic heating and air conditioning ducts supported by the floor should be grounded. Reflective insulation should not be installed where metallic heating and air conditioning ducts supported by the floor are not grounded. The installer should measure the width of the spacing between floor joists to determine the correct width of the insulation required to meet the manufacturer's installation instructions.

The installer should measure the length of runs between obstructions and cut the insulation in a manner which will minimize the number of joints. When installing reflective insulation beneath floors over crawl spaces, the ground surface should be covered with a vapor barrier.

Crawl spaces should have a free ventilation area of one square foot for every 1,500 square feet of the ground area before insulation is applied. These spaces should be cross ventilated where possible.

General: Reflective surfaces should be free of dirt, oil films, and other surface coating before installation. The insulation should be mounted in a flush (see Figure 6.2a) or recessed (see Figure 6.2b) application depending upon the space available. Fasteners of aluminum, stainless steel or plastic should be used.

Figure 6.2: Reflective Insulation Mounting

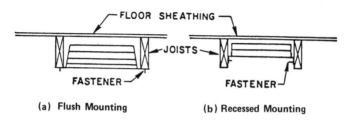

(a) Flush Mounting   (b) Recessed Mounting

Source: DOE/CS-0051

ANALYSIS OF THERMAL PERFORMANCE

Polystyrene Foam for Frame-Sheathing

The insulated frame-sheathing composite insulating assemblies discussed to this point are only applicable to new construction. A variation of the composite assemblies discussed in Chapter 4 is applicable to retrofit construction. Figure 6.3 shows this application as described by the manufacturer of extruded polystyrene foam. Analysis of conductive thermal performance is shown in Table 6.3. As can be seen in this table, there can be a significant improvement in thermal performance. This assembly is most economical only when a decision has already been made by the homeowner to replace existing siding.

Figure 6.3: Extruded Polystyrene Foam Construction

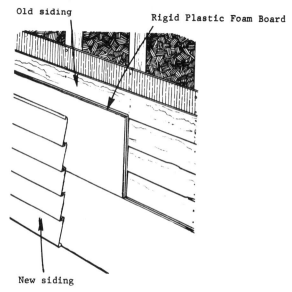

Old siding

Rigid Plastic Foam Board

New siding

Source: BNL-50862

Table 6.3: Comparison of Wall System "R" Values Before and After Retrofit with Extruded Polystyrene Foam

Components of Wall	Wall with No Cavity Insulation		Wall with 1½" Cavity Insulation		Wall with 2½" Cavity Insulation	
	Through Framing	Between Framing	Through Framing	Between Framing	Through Framing	Between Framing
Old Section						
Outside air film	0.17	0.17	0.17	0.17	0.17	0.17
Siding (average)*	0.40	0.40	0.40	0.40	0.40	0.40
Building paper	0.06	0.06	0.06	0.06	0.06	0.06

(continued)

Table 6.3: (continued)

Components of Wall	Wall with No Cavity Insulation Through Framing	Between Framing	Wall with 1½" Cavity Insulation Through Framing	Between Framing	Wall with 2½" Cavity Insulation Through Framing	Between Framing
¾" wood sheathing	0.98	0.98	0.98	0.98	0.98	0.98
3½" wood frame	4.35	—	4.35	—	4.35	—
3½" air space	—	0.97	—	—	—	—
1½" fiber glass batt	—	—	—	4.00	—	—
2" air space	—	—	—	0.94	—	—
2½" fiber glass batt	—	—	—	—	—	7.00
1" air space	—	—	—	—	—	0.92
¾" wood lath and plaster	0.41	0.41	0.41	0.41	0.41	0.41
Inside air film	0.68	0.68	0.68	0.68	0.68	0.68
R value of old section	7.05	3.67	7.05	7.64	7.05	10.62
U value of old section	0.1418	0.2725	0.1418	0.1309	0.1418	0.0942
Framing correction	x 20%	x 80%	x 20%	x 80%	x 20%	x 80%
	0.0284 + 0.2180		0.0284 + 0.1047		0.0284 + 0.0754	
Average U of old wall	0.2464		0.1331		0.1038	
Average R of old wall	4.058		7.513		9.634	

Retrofit Section

¾" extruded polystyrene foam	3.75	3.75	3.75	3.75	3.75	3.75
Aluminum siding (hollow back)	0.61	0.61	0.61	0.61	0.61	0.61
R value of retrofit section	11.41	8.03	11.41	12.00	11.41	14.98
U value of retrofit section	0.0876	0.1245	0.0876	0.0833	0.0876	0.0668
Framing correction	x 20%	x 80%	x 20%	x 80%	x 20%	x 80%
	0.0175 + 0.0996		0.0175 + 0.0666		0.0175 + 0.0534	
Average U of retrofit section	0.1171		0.0841		0.0709	
Average R of retrofit section	8.54		11.88		14.10	
U value improvement	0.1293 (52.5%)		0.0490 (36.8%)		0.0329 (31.7%)	

*Thermal resistance of 0.40 for siding is the average thermal resistance of wood (R = 0.85), asbestos (R = 0.21), and asphalt roll siding (R = 0.15).

Source: BNL-50862

BUILDING CODES AND STANDARDS

The material in this chapter was excerpted from a report prepared by Brookhaven National Laboratory with the assistance of Dynatech R/D Company (BNL-50862).

HISTORICAL PERSPECTIVE

Building codes have served traditionally to protect the safety and health of building occupants and the community. They have existed in cities for nearly four thousand years with the earliest known example being in the Code of Hammurabi in about 1750 BC. The earliest codes were specification codes which defined what materials and methods of construction were acceptable and that type of code has existed to the present time. A performance code defines the desired end result, but does not specify the means of achieving that result. The performance code permits use of new materials and designs as they become available and are found acceptable. Such provisions are essential for an expanding technology and must be included to enable use of improved materials and methods.

The building codes apply to new structures and traditionally have not governed structures erected prior to enactment of the codes. Only when existing structures are extensively renovated have they been required to conform to current building codes. Fire codes have existed for the purpose of controlling community or occupant safety or limitation of property damage. They must be complied with upon construction and also during the life of the structure.

Insurance companies have exerted a strong influence on building design, but traditionally have been concerned with reduction of property damage and consequent lower damage claims. The control of structures by insurance companies has been by means of the premium rate structure with higher premiums required for buildings which loss experience indicates may be subject to greater damage due to fire.

Other controls have existed through limitation of the availability or cost of mortgage money for structures which do not meet minimum requirements. A prime example of this type of control is the Minimum Property Standards of the Federal Housing Administration. Properties meeting those requirements have been eligible for lower interest rates.

Historically, the control of energy efficiency of structures has not been part of the restrictions of building codes; rather the responsibility of codes has been safety. It is a recent concept due to the limitations of energy. For energy use control to become an integral and enforceable part of the building codes, a concensus must be developed among the interested parties in the industry concerning what limitations are desirable, acceptable, cost effective, and energy effective.

ASHRAE STANDARD 90-75

The publication of the ASHRAE Standard 90-75 (64) has provided the basis for energy conservation requirements in many building codes. The model codes developed by the various building code organizations in the United States have based their requirements upon that document.

Standard 90-75 has received criticism and is being revised. Many comments relative to the issue that the minimum requirements for building envelopes are not sufficiently restrictive or are overly restrictive have been advanced. A practical problem with performance standards in general is the inability to monitor or assure compliance, particularly when the number and variations of buildings are as large as in the case of residential or commercial/industrial construction. The value of the standard is that it exists; that is, it specifies minimum design requirements more restrictive than those that previously existed, that it permits design options by trade-offs, and that it has led to incorporation of energy conservation requirements in building codes and other related specifications. Now that the framework of this standard exists, it is much easier to change or upgrade the requirements within that framework. A proper place to work for upgrading is within the American Society of Heating, Refrigerating, and Air-Conditioning Engineers.

ASHRAE Standard 90-75 is a comprehensive component performance document that includes all forms of energy consumed within a building. It relates to heating, ventilation, air-conditioning, lighting, service water heating, and electrical power distribution. With respect to building envelopes, it is based on calculations of steady-state conditions with empirical adjustment for air leakage, solar factor as influenced by latitude, and shading coefficient of the fenestration. For cooling only, the mass of wall construction is considered. However, the actual influence of these factors needs to be demonstrated by further laboratory tests or instrumentation of existing structures. The average thermal transmittance of the gross wall area for different building types as required by ASHRAE Standard 90-75 is given in Figures 7.1 through 7.6.

ASHRAE has recently added a Section 12 to Standard 90-75. This section deals with selection of energy sources and conservation of depletable or nonrenewable energy source.

Figure 7.1: U₀ Walls—Type A Buildings

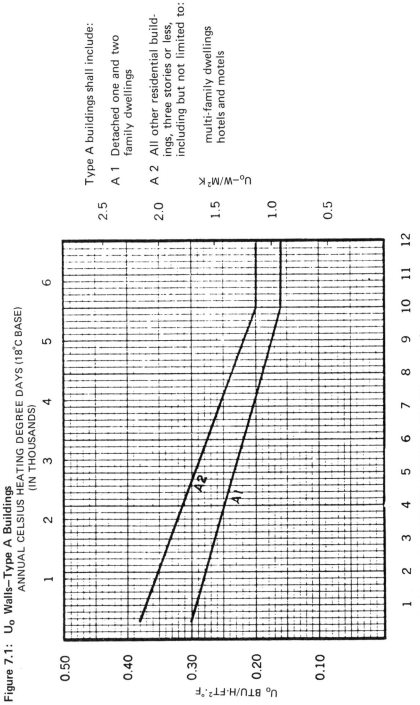

Type A buildings shall include:

A 1 Detached one and two
      family dwellings

A 2 All other residential build-
      ings, three stories or less,
      including but not limited to:

      multi-family dwellings
      hotels and motels

Source: BNL-50862

Figure 7.2: R Values—Slab on Grade

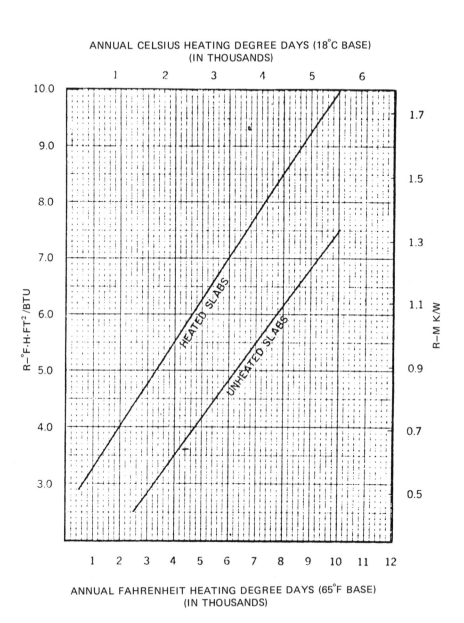

ANNUAL CELSIUS HEATING DEGREE DAYS (18°C BASE)
(IN THOUSANDS)

ANNUAL FAHRENHEIT HEATING DEGREE DAYS (65°F BASE)
(IN THOUSANDS)

Source: BNL-50862

Figure 7.3: U₀ Walls—Type B Building

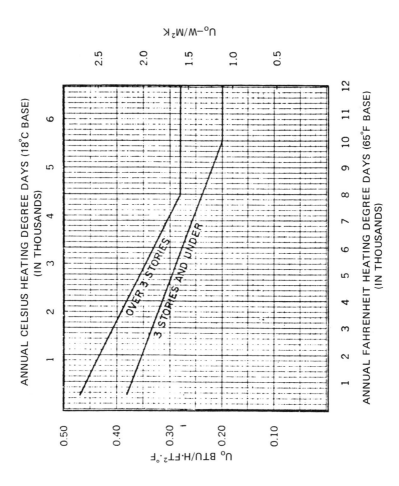

Source: BNL-50862

Figure 7.4: U₀ Roofs and Ceilings—Type B Buildings

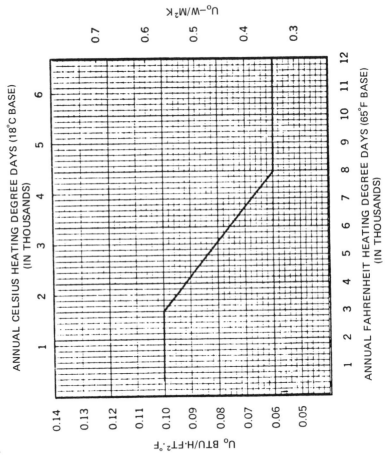

Source: BNL-50862

Figure 7.5: U_o—Floors Over Unheated Spaces

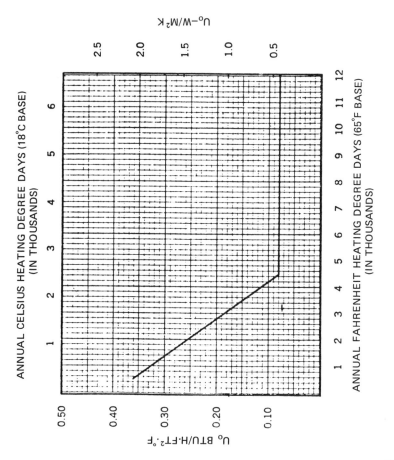

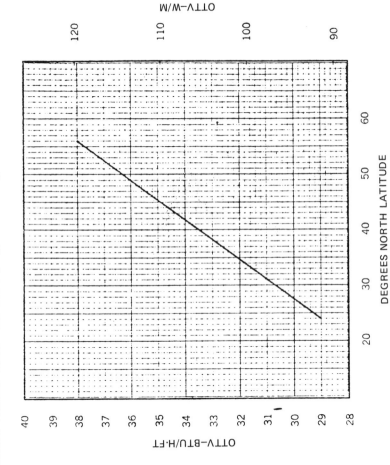

Figure 7.6: Overall Thermal Transfer Value Walls Cooling Type B Building

Source: BNL-50862

In addition to ASHRAE 90-75, which applies only to new buildings, ASHRAE is issuing draft standards for comment on energy conservation in existing structures. The proposed standards are listed below:

> 100.1P–Energy Conservation in Existing Buildings
> Low Rise Residentials (This includes mobile homes).

> 100.2P–Energy Conservation in Existing Buildings
> High Rise Residentials.

> 100.3P–Energy Conservation in Existing Buildings
> Commercial.

> 100.4P–Energy Conservation in Existing Buildings
> Industrial.

> 100.5P–Energy Conservation in Existing Buildings
> Institutional.

> 100.6P–Energy Conservation in Existing Buildings
> Public Assemblies.

Standard 100.6P was released for review in September 1977. The 100.1P requirements are less restrictive than 90-75, while the others are comparable. Standards 100.2P and 100.3P were also issued for comment by the American National Standards Institute. If these standards have the impact of 90-75, inclusion in Building Codes or other regulatory documents may then be anticipated; but enforcement of regulations for existing buildings will require additional legislation or regulation.

THE MODEL CODES

Due in part to the efforts of the National Conference of States on Building Codes and Standards (NCSBCS), the three Model Codes (Basic Building Code, Standard Building Code, and Uniform Building Code) now contain provisions on energy conservation. NCSBCS has been very active in the energy conservation area and has in preparation a draft Model Code for Energy Conservation. A resolution has been adopted by NCSBCS for States to adopt that code.

The Uniform Building Code is administered by the International Conference of Building Officials in Whittier, California. The Uniform Building Code, 1977 Supplement, has added Appendix Chapter 53 on Energy Conservation for all new buildings that are heated or cooled. This is a chapter that rather faithfully copies ASHRAE Standard 90-75 for new construction and includes provisions for exterior envelopes, heating, and piping, lighting and electrical distribution systems. This Appendix Chapter has to be adopted separately by governmental enforcement bodies, but inclusion as a Chapter will lead to the later inclusion into the body of the code.

The Basic Building Code is administered by the Building Officials and Code Administrators International, Inc. in Chicago, Illinois. In 1977 they issued the first edition of *The BOCA Basic Energy Conservation Code* which compiles energy conservation requirements published in the Basic Building, Mechanical, and Plumbing Codes as amended in the 1976 Supplements to the 1975 codes.

These govern "all buildings and structures hereafter erected that provide facilities or shelter for human occupancy." That code copies the intent of Standard 90-75 with minor changes such as use of tables in place of graphs. It permits a "U_o" (combined thermal transmittance value for the gross area of roof ceilings) of 0.06 Btu/hr-ft^2-°F for 8,000 degree days or more, while Standard 90-75 and the Uniform Code require a maximum of 0.05 Btu/hr-ft^2-°F. It does not include provision for the Solar Factor, which corrects for the latitude of the site as provided by Standard 90-75, nor the shading coefficient under cooling criteria for walls. Neither is provision made for the effect of mass. As an amendment, it must be adopted separately.

The 1977 Revision to the 1976 Edition of the Standard Building Code published by the Southern Building Code Congress International, Inc. in Birmingham, Alabama completely revises Appendix "J" on Thermal Performance. The 1976 version covered only Business, Schools, Institutional, and Assembly Buildings, while the 1977 version covers all buildings.

The Code states applicable provisions of ASHRAE Standard 90-75 shall be complied with to meet the requirements of the Appendix. This Appendix must, however, be specifically adopted by governing bodies in addition to the Chapters of the Code in order to be enforceable.

The provisions of Appendix "J" in general follow Standard 90-75, including the figures, but with some important omissions and changes. Not present is the formula for Overall Thermal Transfer Value (OTTV) for mechanically cooled structures. It is stated that the Portland Cement Association 1976 publication *Simplified Thermal Design of Building Envelopes for Use with ASHRAE Standard 90-75* is an acceptable calculation procedure for determining thermal capacitance effects of insulation.

The National Building Code 1976 published by the American Insurance Association in New York City has nothing on energy conservation. Work is underway to include provisions of Standard 90-75 as a supplement or in the next revision. The *One and Two Family Dwelling Code* published jointly by BOCA, AIA, ICBO, and SNCC in 1975 has nothing on energy conservation. Neither does the 1976 supplement.

The building trade journal *House and Home* in the January 1977 issue reported on the existence of State Building Codes as shown in Table 7.1. Upon adoption of the 1977 Basic, Standard, and Uniform Code supplements, the listed States, and possibly others, should have active energy conservation code requirements. Effective January 1, 1978, the State of New Jersey adopted the BOCA Basic Energy Conservation Code/1977.

Table 7.1: Statewide Mandatory Building Codes

. . . . . State	. . Code Type . .
Alaska	Uniform
Connecticut	BOCA
Idaho	Uniform
Indiana	Uniform
Massachusetts	BOCA

(continued)

State	Code Type
Michigan	
(except Detroit)	BOCA
Minnesota	Uniform
Montana	Uniform
New Jersey	BOCA
New Mexico	Uniform
North Carolina	State Written
Ohio	State Written
Oregon	Uniform
Rhode Island	BOCA
Virginia	BOCA
Washington	Uniform
Wisconsin	State Written

Statewide Voluntary Building Codes

State	Code Type
Georgia	Standard
Maryland	BOCA
New York	State Written

Source: BNL-50862

PROPERTY STANDARDS

The Minimum Property Standards of the Federal Housing Administration have great influence on construction. In the July 1974 Revision No. 1 to Minimum Property Standards for One and Two Family Dwellings, Section 607-3 on Building Insulation sets maximum "U" values for building elements. These values do not include adjustments for structural framing nor sash frame or glass door frames.

The Flat Roof Deck maximum requirements are higher (or less restrictive) than ASHRAE Standard 90-75. For example, 4,500 degree days or less sets a maximum "U" of 0.14 Btu/hr-ft^2-°F, while Standard 90-75 sets 0.09. Degree days of 8,001 or more require 0.09 versus 0.06. Residential building walls require a maximum of 0.08 for all areas, while Standard 90-75 requires 0.16 for "U_o" on the gross area including windows and sash for degree days above 10,000. No limit on area of glazing exists although above 4,500 degree days a maximum "U" of 0.65 is required (thermal windows). As an example, a 30' x 20' single story dwelling with 9' walls, 12% thermal glazing, and two doors calculates to a "U_o" (average thermal transmittance of gross area) of 0.16 or the same value as required by Standard 90-75.

The *Federal Register* dated June 17, 1977 published a revision to the HUD Minimum Property Standards for One and Two Family Dwellings effective May 1, 1977. That revision permits "optional use of Standard 90-75 where the results are equivalent to those obtained by using Minimum Property Standards." The published table of maximum "U" values is expanded and made more restrictive, but not quite as restrictive as Standard 90-75. The maximum "U" for thermal windows is increased to 0.69 from 0.65 to permit use of single strength glass. The maximum glass area is limited to 15% except where winter heat gain can be demonstrated to exceed the 25 hour heat loss. It should be noted that the Model Codes, in 1977, reduced the minimum requirement for windows to 10%.

The Farmers Home Administration of the Department of Agriculture published thermal performance standards for Rural Housing loan funds for new construction and resold structures in the *Federal Register*, January 23, 1978 to be effective March 15, 1978. For new dwellings, the maximum "U" values not corrected for framing are more stringent than the cited HUD standards published June 17, 1977. For resale dwellings the same standards apply, but a note allows deviations for walls if attainment of the standard is not economical. No mention is made of trade-offs to attain overall efficiency.

Title 24 issued by the Department of Housing and Urban Development was printed as Part 280 "Mobile Home Construction and Safety Standards" in the *Federal Register* on December 18, 1975, to be effective June 15, 1976. In Subpart F, Thermal Protection, the requirements for Heat Loss/Heat Gain are specified. The country is divided into three winter design temperature zones. Zone 1, going from west to east includes California, Arizona, New Mexico, Oklahoma, Arkansas, Tennessee, and North Carolina. Zone 2 includes the rest of the contiguous 48 states, while Alaska is in Zone 3. The other outlying possessions are included in Zone 1.

The standard states that the transmission heat loss coefficients shall not exceed the following values:

Zone	Maximum $Btu/hr\text{-}ft^2\text{-}°F$
1	0.157
2	0.126
3	0.104

Zones 2 and 3 require storm windows or insulating glass. Infiltration heat loss is limited to 0.7 Btu/hr for each foot of perimeter of the pressure envelope. Transmission heat gains shall be similar. Calculations must be in accord with the 1972 *ASHRAE Handbook of Fundamentals*. Comfort Heating and Cooling Certificates must be affixed to certify the specific design zones which qualified. These requirements are somewhat less restrictive than ASHRAE Standard 90-75 and essentially parallel to National Fire Protection Association (NFPA) Standard 501B, 1977.

CODES AND REGULATIONS OUTSIDE THE U.S.

Some mention should be made of the activities in the Codes and Regulations Area in countries outside the U.S. With the question of the energy crisis becoming increasingly important internationally and with the involvement of the U.S. in the new ISO 163 Thermal Insulation Committee, it is possible that future codes in the U.S., where appropriate, may reflect the philosphy and/or substance of those in other countries. Clearly thermal performance is a prime subject for such consideration.

Since the early seventies considerable activity has taken place, in many European countries in particular, to produce national standards or codes. There has been a trend away from documents covering only the thermal characterization of opaque walls towards those containing more general and functional requirements attempting to cover some of the other important parameters which affect energy consumption for buildings. In a majority of cases these have been concerned only with the more important heating mode or winter condition. Attempts are now being made, particularly within the European Economic Community, to try to bring more conformity and concord between the various national documents (65).

France has been a leader in this area with the implementation on July 1, 1975 of Decree No. 74306 dated April 10, 1974. The United Kingdom implemented Building Regulation 1974, Part F, "Thermal Insulation" on January 31, 1975, while Belgium with Standard NBNB62–001 and West Germany with an addendum to DIN 4108 were also quick to follow suit. Italy and Denmark drafted new national standards while The Netherlands revised NEN1068. The other Scandinavian countries, particularly Sweden, undertook drastic upgrading of their national documents in this field in 1975–76.

In general, the codes are based upon some kind of coefficient or transmission index which limits the maximum quantity of energy which may be lost per unit volume of dwelling per hour per degree of temperature difference between the outside and inside.

Example: The French Document—The overall heat losses (by transmission and infiltration or ventilation) for seven classes of dwellings are limited by the stipulation of a maximum value for a coefficient G for three climatic zones. G is defined as:

$$G \ (W/m^3 K) = \frac{\text{losses of a dwelling per unit temperature difference}}{\text{volume of dwelling}}$$

and $G = G_{tr} + G_{vent}$

A single value of $G_{vent} = 0.34$ W/m^{3}K is being used where the number of air renewals is unknown. This value corresponds to a constant renewal of air equal to the total volume of the dwelling. Using this value, G_{tr} can be derived.

However, the French document is governed further by the application of Regles Th issued in 1975 which provides means of calculating heat losses in structures. In addition it introduces compulsory use of various multiplication factors to modify measured thermal properties of materials or component systems or the calculated values based on measurements on laboratory samples of individual materials. These factors reduce the measured performance to take account of such parameters as effects of moisture vapor and water, air infiltration, etc. This is a very different philosophy to that which is current in the U.S.

It should be pointed out that the use of these factors is becoming more prevalent. In West Germany for example, the test methods for materials and systems themselves contain requirements which are most conservative to adjust the actual measured values for application purposes. This is a trend which should not become general practice.

Certainly it is correct to anticipate actual conditions of application when calculating thermal performance or applying laboratory test results.

However, more definitive information than is available at present is required concerning the real effects of the various parameters on the insulation performances in order to make these correction factors applicable. This is a subject which is relevant to methods of calculation of thermal performance. In this context, at the recent ISO 163 Subcommittee 1 on Test Methods in West Berlin the U.S. delegation, with some of the other delegations, were successful in persuading the whole group that the inclusion of these factors was in fact under the jurisdiction of Subcommittee 2 on Calculation Methods.

Overall, these activities outside the U.S. must in the future have a bearing on the thoughts, considerations, and actions of such organizations as ASHRAE, ASTM and, hence, on the Federal, State and Official Code Organizations. However, the philosophy described should not be adopted without further information being made available to justify any coefficients or indexes used.

BUILDING CODE IMPACTS OTHER THAN THERMAL PERFORMANCE

In addition to specification of permissible heat loss or gain from buildings, the Building Codes influence the use and manner of installation of insulations. While many insulations have desirable properties from a thermal performance aspect some have possible drawbacks due to other considerations. As an example, some cellular plastics have a high thermal resistance per unit thickness and serve as an air barrier to reduce heat loss due to infiltration. However, among the less desirable properties of some foams are fire spread characteristics, emission of large quantities of smoke during combustion, possible emission of toxic fumes during combustion, and deterioration of properties with time.

To protect building occupants from the hazards associated with the above fire properties, codes have set maximum surface flame spread and smoke developed requirements for foams as well as other products. They also require that foams not be installed with exposure to habitable spaces. Appropriate fire barriers such as ½" gypsum board must be installed between habitable spaces and foams. Such thermal protection delays the time of involvement of plastic foams in an interior fire and gains additional time for evacuation of the structure during a fire.

The toxicity of smoke and fumes emitted from building structures during fires is a matter of increasing concern. All organic materials emit smoke and various gases during combustion. Carbon monoxide is a component of all fire gases and is recognized as one of the major causes of fatalities. Exposure of building inhabitants and fire fighters to products of combustion could result in long-term harmful effects and even death. Those active in the study of fire and in building codes have done a great deal of investigation into toxicity of materials during fire, but the subject is very complex and no satisfactory means of measurement has yet been devised.

In response to the need for regulation, all of the Model Codes had requirements that emitted gases must be no more toxic than those emitted by wood burned under similar conditions. This requirement has not been possible to enforce due to lack of an acceptable, statistically significant method of test, and therefore has been removed.

The Committee of the American Society for Testing and Materials which is responsible for Fire Tests of Materials and Constructions has studied the problem for several years and will have a Recommended Practice for Biological Evaluation of the Toxic Effects of Smoke. They also will draft a test method aimed at identifying unusually toxic materials. Upon acceptance of such test methods, it is probable that Building Codes will include restrictions on materials, including insulations, based on accepted test standards.

CONCLUSIONS

Codes and specifications appear to reflect adequately the state-of-the-art and consensus standards with regard to thermal insulation. As knowledge of materials, testing and systems increases, it is important that the codes keep abreast of developments and update those sections dealing with thermal insulations.

An existing deficiency in 1977 was that the Model Codes had only appendixes that referred to energy conservation. This was inherent in the method of operation of the code bodies. It is, therefore, of major importance that governmental bodies adopt these appendixes as part of their local codes so the provisions for energy conservation for new buildings will have the force of law.

It is suggested that the National Conference of States on Building Codes and Standards continue their efforts to induce states, municipalities, and federal agencies to adopt energy conservation code provisions.

Upon adoption, the problem will be code enforcement. Building officials and inspectors may not be sufficiently familar with insulation theory and practice for enforcement as thermal performance of insulations has not been a part of building codes. The model code agencies have traditionally conducted courses for building officials and inspectors on code interpretations.

A start was made on such education in the energy conservation field. An Energy Conservation Summer Institute was conducted at Salt Lake City, Utah during August, 1977. Programs for building inspectors, plan reviewers, and building officials were included. This program was sponsored by the Model Code organizations under contract with ERDA. The Building Officials and Code Administrators International (Administrators of the Basic Code) announced a three day program on energy conservation presented at the BOCA meeting in January, 1978.

Assuming the ASHRAE recommendations for refitting of existing buildings (drafts 100.2P through 100.6P) are accepted, a problem of enforcement will arise. Changes in building codes have not, for the major part, been retroactive. To adopt and enforce such provisions on refitting will require a change in philosophy and possible legal tests of constitutionality. This is a procedural matter that must be examined carefully to determine legality and public acceptance. If found acceptable it may be decided that enforcement should be by means other than building codes.

PART II

INSULATION
FOR INDUSTRIAL PROCESSES
(Nonstructural)

IMPORTANCE
OF INDUSTRIAL INSULATION

The material in this chapter was excerpted from reports prepared
by York Research Corporation (PB-259 937) and by Oak Ridge
National Laboratory (TID-27120).

INTRODUCTION

Industrial plants and utilities account for about half of the total United States
energy use. Historically, this sector has utilized thermal insulation to protect
personnel, maintain process temperatures and conserve energy. In this period
of rising fuel costs, thermal insulation is probably the best proven and univer-
sally applicable solution available to industry for the conservation of this expensive commodity.

In past years, when first-cost analysis was the primary criterion for capital expenditures, using more than minimal amounts of insulation was not justified.
This rationale is no longer acceptable in the present era of expensive energy.
Consequently life cycle costing, especially in areas of energy conservation, is
now becoming the predominant industrial design criterion. Insulation specified
around today's economics saves on the average 30 to 40% more energy than is
being conserved with the outdated insulation designs of the past. This is not
only a significant factor in new plant construction, but also provides the financial justification for retrofitting insulation in existing facilities.

The conservation of energy through the use of optimal economic insulation
thickness has obvious benefits for industry, and equally impressive potential
benefits for the United States. It has been estimated that over 1,400 trillion
Btu, the equivalent of over 122 million barrels of oil could have been saved
in 1974, had industry alone installed economic insulation thicknesses. This is
in addition to the energy saved with existing insulation. The potential additional energy conservation available to industry through the use of economic
insulation through 1990 is estimated to be the equivalent of 3.5 billion barrels
of oil or 250 million barrels per year. An extra benefit to society is a reduction

in air pollution, which necessarily follows decreased fuel use. As with any capital expenditure, the dollars spent for insulation are expected to provide the plant owner with a certain return on his investment. (This return can be in the form of personnel and equipment protection as well as in energy savings.)

In addition to economic insulation thickness, the following topics are also important: available insulation materials for cryogenic through high temperature applications; energy conservation potential for insulation in various industrial applications; economic considerations for application of insulation; availability of thermal property data; and thermal property testing and specification of insulation materials.

POTENTIAL ENERGY SAVINGS

The potential for energy savings through effective use of insulation was estimated to be about 1.5 x 10^{15} Btu (1.5 quads) per year for the six largest energy consuming industries (see Table 9.5, page 160). Such savings would reduce the total process energy consumption of these industries about 9% and would reduce our nations's total energy consumption by 2%.

Of the total potential energy savings identified, approximately 50% can be realized through increased insulation and maintenance of steam distribution systems used through industry (see Table 9.6, page 160).

Other areas for application of insulation include skid rails in metal reheating furnaces, petrochemical reaction vessels, glass melting furnaces, cement kilns and food processing equipment.

Barriers to the more effective utilization of insulation include: unawareness of the energy/cost benefits; uncertainty in long-term performance of insulation; and inability to properly specify the requirements of efficient insulation systems.

SUMMARY OF TECHNOLOGY ASSESSMENT

Materials Design and Selection

A large variety of thermal insulation materials are manufactured for application in the various industrial temperature ranges and environments. In many cases, materials design and selection are based on properties other than thermal resistance (see Table 9.2, page 150).

Although adequate materials can be found for many applications, increasing demands by users for long-term systems performance will require materials having improved and optimized properties. Included in this context are materials not only for the primary thermal resistance but also, jacketing, sealing and joining materials, as well as improved methods for installing and joining these materials. To accomplish such improvements, close communication and cooperation between insulation users, contractors, and manufacturers are required.

In some cases, insulation systems are overdesigned to prolong satisfactory performance; system performance data are lacking for most applications. Over-design of systems results in higher investment costs and, therefore, represents a deterrent to increased usage of insulation.

Economic Considerations

Economic considerations affecting decisions to install insulation include background factors, validity and confidence in analysis methods for insulation system design, and economic/business constraints placed on capital investment decisions. The direct value of insulation is based on projected savings in fuel costs, increased productivity and quality of product, availability of fuel forms (e.g., natural gas), and compliance with environmental and personnel safety requirements. Shortages and cost of investment capital deter some companies from upgrading existing insulation systems, although thermal insulation is recognized for its relatively high rate of return on investment.

Thermal Property Data

The assessment of thermal property data has been approached from several viewpoints: availability, validity, and relationship between available data and insulation system performance. The large volume of thermal data that have been generated over the years have been provided primarily by a relatively few major manufacturers of insulation materials. It has been indicated that portions of these data remain unpublished (and in some cases are company proprietaries).

Published data are fragmented in various trade journals and sales literature of the respective companies. Data given in handbooks are often presented as typical or for a generic material without indicating a range of properties available for specific commercial products. Very little of the data has been developed by independent organizations or within governmental facilities. The capacity to develop such data within governmental facilities is generally lacking.

Opinions expressed by insulation users on the validity of reported property data ranged from complete satisfaction to total distrust. Reasons given for distrust of data include:

(1) The data are obtained usually under controlled laboratory conditions on selected samples. Typically the materials are tested dry and only over a narrow temperature range,

(2) recommended standard test procedures (e.g., American Society for Testing Materials) are not sufficiently definitive, test results are dependent to some (but unknown) degree on the expertise of the operator, and,

(3) limited comparisons of test results between various facilities have shown significant ranges in the data. Standard reference materials for interfacility calibration of testing equipment and procedures are totally lacking, therefore practical means for rectifying measurement differences are not available.

Very little data are available on the long-term performance of insulation materials. Although there is an increasing demand for such data, the availability of these data will be dependent on the development of new test procedures, testing equipment and facilities to perform the tests.

INDUSTRIAL THERMAL
INSULATION MATERIALS

The material in this chapter was excerpted from a report prepared by Oak Ridge National Laboratory (TID-27120).

ASSESSMENT AND DISCUSSION OF AVAILABLE INSULATION MATERIALS

To permit a useful technological assessment of industrial thermal insulation materials, they are conveniently classified on some logical basis related to their use.

The basis chosen for this assessment is the temperature, since the applicability
of any given insulation material is typically limited by an upper temperature at
which it becomes structurally unstable or noncompetitive because of heat transport through it (relatively high thermal conductivity). An indication of the
many types available is shown in Table 9.1.

Most thermal insulation materials also have a lower temperature limit of applicability due to undesirable properties. Several properties of thermal insulations
are important in addition to the thermal conductivity.

Often, some of these other properties will dictate the choice of material rather
than the heat transport characteristics alone. A number of these properties are
identified in Table 9.2. Standard methods for determining several of these properties are not available, and one must therefore be cautious in comparing a specific property of various insulations and using this as the only basis for materials
choice in a given system design. The values of many of these properties for various types of commercially available insulations are unknown. For convenience,
in this assessment the temperature scale is divided into the following ranges related to the range of use of several commercially significant thermal insulations.

(1) Cryogenic: $-270°C (-454°F) \leqslant T \leqslant -100°C (-148°F)$
(2) Low temperature: $-100°C (-148°F) \leqslant T \leqslant 100°C (212°F)$
(3) Intermediate temperature: $100°C (212°F) \leqslant T \leqslant 500°C (932°F)$
(4) High temperature: $T > 500°C (932°F)$

Table 9.1: Industrial Insulations by Temperature Range of Application

Cryogenic and Low Temperature [−270°C (−454°F) ⩽ T ⩽ 100°C (212°F)]

    Evacuated
       Multifoil
       Opacified powders
    Mass Type
       Glass foams
       Organic foams
       Fiber glass
       Loose-fill
       Balsa wood

Intermediate Temperature [100°C (212°F) ⩽ T ⩽ 500°C (932°F)]

    All Inorganic
       Perlite
       Calcium silicate
       Foam glass
       Mineral wool
       Reflective
       Loose-fill
       Insulating firebrick

High Temperature [T > 500°C (932°F)]

    All Inorganic, Carbon, or Metallic
       Loose-fill
       Reflective
       Insulating firebrick
       Ceramic foams
       Ceramic fiber
       Pyrolytic carbon
       Carbon fibers

Source: TID-27120

Table 9.2: Selected Properties of Thermal Insulations Often Used in Design

Property	Rigid and Semirigid	Flex-ible	Blanket, Felt or Batt	Loose Fill	Reflec-tive	Cryo-genic	Insu-lating Fire Brick	Sprayed, Foamed in Place
Thermal conductivity	x	x	x	x	—	—	x	x
Thermal diffusivity	x	x	x	x	—	—	x	x
Maximum temp. limits	x	x	x	x	x	x	x	x
Density	x	x	x	x	x	x	x	x
Abrasion resistance	x	x	—	—	—	—	x	x
Alkalinity	x	x	x	x	x	—	x	x
Capillarity	x	x	x	—	—	—	—	x
Flame spread index	x	x	x	x	—	—	—	x
Smoke density index	x	x	x	x	—	—	—	x
Combustibility	x	x	x	x	—	—	—	x
Compressive strength	x	x	x	—	—	x	x	x
Hygroscopicity	x	x	x	x	—	—	x	x
Linear shrinkage at maximum temp.	x	x	x	—	—	—	x	x

(continued)

Table 9.2: (continued)

| Property | . . . . Insulation Type . . . . | | | | | | | |
	Rigid and Semirigid	Flex- ible	Blanket, Felt or Batt	Loose Fill	Reflec- tive	Cryo- genic	Insu- lating Fire Brick	Sprayed, Foamed in Place
Specific heat	x	x	x	x	x	x	x	x
Tensile strength	x	x	x	—	—	—	x	—
Water absorption during submersion	x	x	—	x	—	—	—	x
Water vapor transmission	x	x	—	—	—	—	—	x
Heat transmission	—	—	—	—	x	x	x	x

Source: TID-27120

Cryogenic Range [-270°C (-454°F) ≤ T ≤ -100°C (-148°F)]

Thermal insulations for use in the cryogenic range consist of both vacuum types and so-called massive insulation, in which one or more solid phases are distributed along with a gas, often dry air, to produce an acceptably low thermal conductivity. Vacuum cryogenic thermal insulation systems typically consist of highly polished metal supporting walls with a vacuum space between them. Within this vacuum space there may be multiple metal reflective foils or various inorganic or organic materials coated with metals (so-called opacified powders) depending upon usage temperature. These insulations as applied to industrial systems are typically custom-designed and installed by the insulation vendors.

Performance of these systems after installation is typically guaranteed by the insulation designer or vendor. Manufacture and improvement of vacuum types of cryogenic insulation are highly sophisticated and very specialized. The multilayer evacuated multifoil insulations have typical thermal conductivities in the range of approximately 3×10^{-5} W m^{-1} K^{-1} (2×10^{-4} Btu in. hr^{-1} ft^{-2} °F^{-1}) to about 2×10^{-4} W m^{-1} K^{-1} (0.0014 Btu in. hr^{-1} ft^{-2} °F^{-1}) and are used in the temperature range from very near 0 K (-459.7°F) up to ambient in some cases.

Other evacuated cryogenic insulations include vacuum spaces containing: (1) opacified powders, (2) glass fibers, or (3) uncoated oxide powders. These three types cover a thermal conductivity range of about 1.7×10^{-3} W m^{-1} K^{-1} (0.0125 Btu in. hr^{-1} ft^{-2} °F^{-1}) to about 0.08 W m^{-1} K^{-1} (0.555 Btu in. hr^{-1} ft^{-2} °F^{-1}), with their thermal conductivities ranking from (1) through (3) in order of increasing conductivity.

Because of the relatively high cost of the evacuated cryogenic insulation system and difficulty of application in some design situations, many types of massive insulations have been examined for use in cryogenic applications. Early materials used for moderately low temperatures included cork and some other natural organic materials. One material, which derived from World War II and which initially showed great promise, was expanded or foamed polyurethane. This material can be foamed into the form of flexible sheets, foamed in place, and foamed into rigid insulation sections. Other foamed thermal insulation materials are polystyrene and rigid foamed glass. Foamed glass is manufactured to size for piping, etc., or can be shaped during field installation. Foamed urethanes, styrenes, and foam glass owe their relatively low thermal conductivities to their microstructure of closed cells containing air or other gas, which results in greatly reduced heat transfer by gas phase convection and conduction. These materials differ considerably in resistance to permeation by water vapor and air, with the result that their service lives in cryogenic applications can be quite different. The basic patents for the foamed glass thermal insulation date from 1934 to 1938 in France, and work in the U.S. on this type of material started in 1937.

Low Temperature Range [-100°C (-148°F) ⩽ T ⩽ 100°C (212°F)]

Thermal insulation materials that have been used in this temperature range indiclude some of the evacuated systems discussed previously as well as many non-evacuated foams, powders, and fibers. The thermal conductivities of these systems range typically from 8×10^{-3} W m^{-1} K^{-1} (0.055 Btu in. hr^{-1} ft^{-2} °F^{-1}) to 0.1 W m^{-1} W m^{-1} K^{-1} (0.7 Btu in. hr^{-1} ft^{-2} °F^{-1}). Foamed glass has a thermal conductivity of about 0.05 W m^{-1} K^{-1} (0.35 Btu in. hr^{-1} ft^{-2} °F^{-1}) at the mid-temperature of this range (32°F), whereas a urethane foam with density of 2 lb/ft^3 (32 kg/m^3) has a reported conductivity value of about 0.02 W m^{-1} K^{-1} (0.15 Btu in. hr^{-1} ft^{-2} °F^{-1}).

However, the conductivity value of some insulation materials employed for cryogenic and low-temperature applications increases dramatically in service because of permeation by water vapor, exchange of the low-conductivity gas originally in the cells by air, and possibly by physical deterioration by ultraviolet light. Caution should be exercised in selecting a material for cryogenic or low-temperature insulation applications unless the water vapor permeation properties are well known, documented, and accepted.

Foamed polystyrene is another insulation material that has been used for various low-temperature applications. For these applications, it suffers from having relatively high moisture permeability, relatively low mechanical strength, high burning rate, and an absolute maximum service temperature of about 80°C (176°F) due to its softening temperature of about 90°C (194°F). This low softening point makes the typically required periodic drying of a low-temperature system insulated with foamed polystyrene somewhat delicate. The relative moisture permeabilities of various low-temperature insulations are shown below (1).

Material	Permeability (perms)
Foam glass	~0
Urethane foam	0.3-6
Polystyrene foam	1-4
Mineral wool	100-200

Foamed plastic insulations are also reported to have a brittle-to-ductile transition at low temperatures. In the temperature range from about -30°C (-22°F) to the upper limit of this range [100°C (212°F)] organic and inorganic massive insulations are used. These include compressed and granulated cork, sandwiched cellular glass and felt board, glass fibers bonded with organic resins, expanded and cellular forms of polystyrene, polyurethane foams, rubber and resin combinations, vinyl chloride cellular foams, wood fibers with suitable binders, polyvinyl acetate, cork-filled mastic, expanded vermiculite and perlite, and aluminum foil on paper. Various loose fill forms of some of these insulation materials are also used. Selection of one of these insulations over others is typically dictated by three factors.

The three factors are: (1) form (i.e., rigid board versus flexible board versus loose-fill) and requirements of the insulation design, (2) price, and (3) availability. Actual trade-off analyses of thermal insulations for use in this temperature range based on accurately known thermal conductivities, moisture permeabilities, and other properties in the service environment are apparently quite rare.

Intermediate Temperature Range [100°C (212°F) ≤ T ≤ 500°C (932°F)]

Thermal insulations available for use in this temperature range include inorganic ceramic materials such as glass fibers, rock wool fibers, various bulk and powdered oxides, calcium silicate, bulk glasses, high- and low-density ceramic bricks, and some specialty reflective systems. This temperature range is very important for industrial applications, as it includes temperatures typically encountered in steam systems and most chemical and petrochemical processes. Many of the insulations used in this temperature range have maximum service temperatures above 500°C, and thus discussion of them relevant to this temperature application range should not be interpreted as indicating maximum use temperature unless referred to specifically.

An important type of insulation that is used extensively in this temperature range is that based on fiber-bonded hydrous calcium silicates. Until recently, asbestos fibers were used extensively to provide calcium silicate insulation with reasonable mechanical strength for use on piping, etc. Glass fibers are now generally employed for this purpose because of health hazards associated with asbestos. Other insulation types used in this range include: (1) diatomaceous silica, (2) cellular glass to about 450°C, (3) glass fiber bonded with high-temperature binders, (4) magnesium carbonate with asbestos or other fibers and binders to about 300°C, (5) rock wool or mineral-derived fibers, (6) expanded perlite with binders, (7) ceramic fibers based on the Al_2O_3-SiO_2 system (aluminosilicate fibers), silica fibers, Al_2O_3 fibers, and ZrO_2 fibers and vermiculite, and (8) metal sheet reflective systems.

The lowest reported thermal conductivities for materials in this group at the mid-range (250°C or about 500°F) are for fiber-bonded calcium silicate, glass or mineral fiber, Al_2O_3-SiO_2 ceramic fibers, diatomaceous silica, and the stainless steel reflective systems. These materials all have similar reported conductivities of about 0.07 W m^{-1} K^{-1} (0.5 Btu in. hr^{-1} ft^{-2} °F^{-1}). All the insulation materials mentioned relevant to this temperature range have thermal conductivities that increase almost linearly with temperature over temperature ranges of 200° or 300°C.

The selection of a particular insulation for use in this temperature range is partially dictated by the value of the thermal conductivity, but is often strongly influenced by other factors such as mechanical properties, forms available, and cost of installation. In the case of multiple metal sheet reflective systems, they are the only acceptable insulation type for nuclear reactor systems because of the requirement for rapid identification and isolation of any steam or water system leaks and accessibility for radioactive decontamination if required. Insulations for reactor systems must also be readily demountable to allow periodic inspection of piping, welds, etc.

High Temperature Range [T > 500°C (932°F)]

Thermal insulation materials available for use in this temperature range become limited in number when the temperature exceeds about 650°C (1200°F), and materials availability becomes quite restricted above about 1100°C (2012°F). Insulations for use above 500°C are all inorganic and in some cases initially consist of hydrated phases that dehydrate and shrink at high service temperatures. For service above 500°C eight principal types of thermal insulation are available. They are (1) calcium-silicate-based material primarily for piping systems, (2) mineral fiber, (3) ceramic fibers based on the Al_2O_3-SiO_2 system, (4) oxide fibers, primarily Al_2O_3 or ZrO_2, (5) carbon fibers, (6) rigid ceramic insulating brick, (7) castable ceramic insulating refractories, and (8) multiple metal foil systems for vacuum applications.

Calcium silicate insulations were discussed in the preceding section. In the case of fibrous insulation applications, a choice between mineral fiber (or rock wool), or carbon and ceramic fibers based on the Al_2O_3-SiO_2 system is based on maximum use temperature, environment, and economic considerations. Until recently, the highest continuous use temperature for any fibrous insulation exclusive of carbon was about 1260°C (2300°F). This is a typical maximum temperature recommendation for ceramic fibers derived from the clay mineral kaolin with a typical SiO_2/Al_2O_3 weight ratio of about 1.13.

The purer single-oxide fibers, which have appeared on the market recently, have much higher reported service temperatures, reportedly exhibit much less dimensional shrinkage near their maximum service temperature than do aluminosilicate fibers, and are more stable in strongly alkaline or acidic environments than aluminosilicate fibers.

Single-oxide refractory fiber insulations are presently based upon either Al_2O_3 or CaO-stabilized ZrO_2, and they are also more stable in reducing atmospheres at high temperature than are fibers that contain silica either as free SiO_2 or in a silicate. The principal deterrent to the wide application of the single-oxide fibers at present appears to be related to (1) their relative newness, (2) their relatively high cost (about 50% more than an aluminosilicate fiber installation of comparable size), and (3) present concerns about the long-term durability of the refractory anchoring systems required to secure the insulation batts in front (hot side) of the other furnace wall components.

These products appear to offer superior performance in those applications requiring their more refractory properties, and it is expected that they will find more widespread use as energy costs continue to increase. It is difficult to specifically compare thermal property values of various fibrous types of insulations, and considerable care must be exercised in doing so because of the different densities of various fibrous products, different resiliencies, and other factors that affect samples used for measurement. Some available thermal conductivity data for Al_2O_3, ZrO_2, and aluminosilicate fibers indicate that above 500°C the conductivity increases with temperature with the conductivity versus temperature curve concave upward, as expected when radiation transport and gas phase conduction are significant.

The thermal conductivity values published for the Al_2O_3 fiber material indicate that it has conduction properties similar to aluminosilicate fiber batts up to

1100°C [k ≈ 0.45 W m^{-1} K^{-1} (3.15 Btu in. hr^{-1} ft^{-2} °F^{-1})] , the highest temperature for which conductivity data are available for the more refractory fiber.
Similarly, the ZrO_2 fiber material at the same density of about 6 lb/ft^3 (10 kg/m^3)
and at 1100°C has a thermal conductivity of about 70% of that of the Al_2O_3
fiber insulation at 1100°C. Thermal conductivity data for these materials at
higher temperatures were not available. Maximum use temperatures of 1595°C
(2900°F) have been claimed for the ZrO_2 fibers and 1510°C (2750°F) for the
Al_2O_3. No thermal conductivity values were located for these relatively new
materials above 1100°C (about 2000°F).

The insulating fiber batt has wide potential application in cyclic furnacing operations or others where the heat capacity of the furnace structure is significant in
determining the energy consumption of the process. Use of insulation batting
as the front or inner wall of these furnaces greatly reduces the amount of heat
input to the furnace required to achieve a given charge temperature. Batts can
also lower heat losses in continuous furnaces in that they have a significantly
lower conductivity than insulating bricks. The reported thermal conductivity
values for Al_2O_3 and ZrO_2 fiber batts with densities of 6 lb/ft^3 (10 kg/m^3) are
significantly lower than those published for the higher temperature grades of
insulating firebrick [i.e., for use above 1260°C (2300°F)] .

At 1093°C (2000°F), available data show that the Al_2O_3 batt material has a conductivity about 85% of that of the insulating firebrick ordinarily used in this
temperature range while the ZrO_2 batt material has a reported conductivity about
45% of that of the insulating firebrick used at its somewhat higher temperature
limit.

Ceramic castable and ramming refractories for use to about 1650°C (3000°F)
have published thermal conductivity values of 1.2 to about 2.2 W m^{-1} K^{-1}
(8–15 Btu in. hr^{-1} ft^{-2} °F^{-1}) at 982°C (1800°F). The compositions of these materials are quite variable, and the manufacturers of these products frequently adjust the compositions for particular applications. These materials have much
higher conductivities than the previously discussed insulating fiber batt structures
or insulating brick and are not very competitive in terms of resistance to heat
transfer with the refractory fiber materials or more conventional insulating firebrick.

These materials do offer a unique solution to many refractory application problems due to the relative ease of installation and relatively wide range of compositions available. Castables and ramming mixes usually have relatively high poros-
ity and therefore their thermal conductivity in service is a strong function of
the ambient gas in the service atmosphere.

Multiple foil or metal sheet reflective insulations for high-temperature use are
usually fabricated of refractory metal such as molybdenum or tungsten separated
by refractory metal pins or spacers. These units can be used only in vacuum,
inert, or reducing atmosphere and are typically employed in electrically heated
furnace equipment having refractory metal heating elements. These insulations
function as multireflective radiation shields and are quite effective for their specialized application at hot-wall temperatures as high as 2500°C (4500°F). Their
cost is relatively high, and like the multifoil evacuated insulations for very low-
temperature cryogenic applications, they are employed because in their particular
application there is essentially no alternative.

Carbon fibers and carbon fiber batts or felt are also effective high-temperature thermal insulations for use in either vacuum or inert-atmosphere environments. These insulations are typically used in furnaces where refractory metal multiple reflecting foil insulation systems are considered too massive, fragile, expensive, or all three. The major considerations concerning use of carbon fiber insulations are the effective carbon pressure at the operating temperature, oxygen pressure in the furnace atmosphere, and potential reactions between the carbon and metal components within the furnace. The equilibrium carbon pressure is relatively high at a given temperature compared with other refractory compounds, and this tendency for carbon evaporation is objectionable in some potential applications of carbon or graphite felt insulations.

Carbon fiber insulations having controlled anisotropy have been marketed in the past few years, and these reportedly have superior thermal insulating properties relative to graphite foil and typical graphite felt materials. Thermal conductivity values reported for rigid oriented carbon fiber insulation vary from 0.04 W m^{-1} K^{-1} (0.3 Btu in. hr^{-1} ft^{-2} °F^{-1}) at about 540°C (1004°F) to 0.19 W m^{-1} K^{-1} (1.4 Btu in. hr^{-1} ft^{-2} °F^{-1}) at about 1430°C (2606°F). The temperature dependence of the conductivity of this material is illustrated in Figure 9.1.

Figure 9.1: Thermal Conductivity of Rigid Fibrous Carbon Insulation*

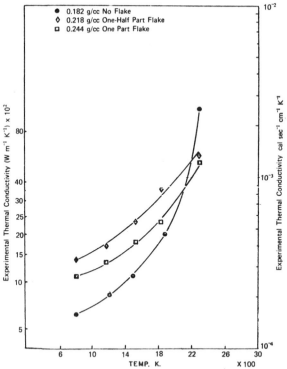

*This shows the effect of graphite flake additions on radiative and conductive heat transmission (1).

Source: TID-27120

At 540°C, the available data show this material to have a conductivity about 40% that of graphite felt at the same temperature. Similarly, it has a conductivity about an order of magnitude less than that of graphite foil at about 600°C. The variety of carbon-base insulations available makes them attractive for many applications. The foils and felts of carbon are essentially as flexible as those made of glass or oxide fibers. The rigid, oriented type may be obtained over a wide range of compressive strengths with the strongest types showing somewhat higher conductivities because more of the carbonized bonding agent is used. An important aspect of these carbon products is that they experience irreversible increases in conductivity with time at high temperatures (>2000°C) because of incipient graphitization of the carbon fibers.

ENERGY CONSERVATION POTENTIAL OF THERMAL INSULATIONS

In this section various process energy requirements for energy-intensive manufacturing industries are analyzed to assess the energy saving potential by use of either more of presently available thermal insulations or better than state-of-the-art insulations. In this assessment the most energy-intensive industries were first identified, then those unit operations that have potential for improved energy conservation were pinpointed, and finally those operations where thermal insulation can play a more significant role were identified.

Several analyses (2)-(8) have recently been conducted of selected industries in which the manufacturing energy input and losses associated with the final product are identified. These efforts have been of considerable help in the identification of the future energy conservation potential of various manufacturing industries and identification of areas where thermal insulations are or could be important. The estimated accuracies of the energy figures found were usually no better than ±20%.

The industries that use most energy in the industrial sector are (1) steel, (2) aluminum, (3) chemical, (4) petroleum, (5) paper, (6) stone-clay-glass-cement, and (7) food. The energy consumption of the significant industries in periods for which data could be identified (8) are given in Table 9.3.

Table 9.3: Energy Consumption of Selected Industries

| Industry | ...Energy Consumption* ... | | Period |
	10^{18} J	10^{15} Btu	
Aluminum	1.5	1.4	1974
Steel	3.6	3.5	1972
Other metals	1.1	1.0	1972
Chemical	4.9	4.6	1973
Petroleum	3.2	3.0	1971
Paper	2.7	2.6	1972
Glass	0.3	0.3	1972
Structural clay products, cement and others	2.1	2.0	1972
Food	1.4	1.3	1971

*These energy values were computed on the basis that electricity, where employed, cost an energy equivalent of 9,935 Btu/kWh for an assumed average electric power plant efficiency of 34%.

Source: TID-27120

These seven industries were estimated to consume about 75% of the total energy requirements of all manufacturing industries in 1975.

The amount of energy rejected as heat in various temperature ranges per year for these manufacturing industries has been estimated by Reding and Shepard (8) and these will be discussed since some of this rejected heat represents energy potentially conservable through better use of thermal insulations. Some of Reding and Shepard's data are shown in Table 9.4 for the six subject industries.

Table 9.4: Estimated Heat Rejection in the Six Biggest Fuel Consuming Industries (1972–1973)

Industry (or Process) and Operation	Radiation, Convection, Conduction and Unaccounted Losses**	. . . . Rejected Heat (10^{12} Btu/yr)*				
		Below $100°C$	$100°-250°C$	$250°-800°C$	$800°-1800°C$	Total All Ranges
Chemical						
Chlorine/caustic soda	40	254	51	–	–	–
Ethylene/propylene	20	151	79	40	–	–
Ammonia	32	115	139	40	–	–
Ethylbenzene/styrene	1	24	28	–	–	–
Carbon black	2	–	32	–	–	–
Sodium carbonate	2	36	16	–	–	–
Oxygen/nitrogen	1	171	147	–	–	–
Cumene	–	4	4	–	–	–
Phenol/acetone	–	8	8	–	–	–
Other	150	628	290	79	–	–
Total	248	1,391	794	159	–	2,345
Primary Metals						
Steel	298	79	397	596	795	–
Aluminum	119	378	79	20	79	–
Other	119	397	119	40	79	–
Total	536	854	595	656	953	3,058
Petroleum	318	994	497	695	–	2,186
Paper	199	1,391	755	119	–	2,265
Glass—Cement—Other						
Cement	119	79	40	219	–	–
Glass	99	60	12	87	36	–
Other	99	79	40	159	20	–
Total	317	218	92	465	56	1,049
Food	60	795	159	119	–	1,073
Grand Total	1,679	5,643	3,110	2,213	1,009	11,975
Ratio to Grand Total		0.47	0.26	0.18	0.09	

*To convert to 10^{15} J/hr, multiply by 1.0543. The estimated accuracy of the values is ±20%.
**Radiation, etc. losses are distributed within other losses shown in the table.

Source: TID-27120

Analysis of the data in Table 9.4 provides a reasonable estimate of the total rejected energy in the form of heat from each of the major industries in each temperature range. The total rejected heat is approximately 12.0×10^{15} Btu (12.7×10^{15} J), and this represents about 56% of all the energy used by these industries per year in the 1971-1973 period. The grand totals in Table 9.4 indicate the estimated quantities of energy in the form of heat available in the various temperature ranges. These estimates show that a very large fraction (47%) of the rejected heat is available at a temperature less than 100°C. Similarly, about three-fourths of the rejected heat is available at less than 250°C and about 27% of the rejected heat is available in the temperature range 250° to 1800°C.

The biggest potential for immediate-term energy savings is thus obviously in those manufacturing process operations in which relatively low-quality heat is rejected (i.e., at temperatures below 250°C). This loss is about 8.8×10^{15} Btu/year (9.3×10^{15} J/year). Similarly, there is a significant potential for immediate-term energy savings in manufacturing process operations where the rejected heat is available above 250°C. About 3.2×10^{15} Btu (3.4×10^{18} J) is rejected in this higher temperature range. In terms of crude oil equivalent, 8.8×10^{15} Btu/year (9.3×10^{18} J/year) represents about 1.5×10^9 barrels, assuming (9) a barrel of crude oil produces 5.8×10^6 Btu (6.1 GJ). This represents about 58% of present U.S. crude oil imports.

Also, 3.2×10^{15} Btu is roughly equivalent to 21% of the present U.S. crude oil imports. Roughly about a third of the total conservable energy in these industries might be conserved via use of more thermal insulation and by use of improved insulation maintenance. Thus, appreciable energy conservation via better thermal insulations and other techniques could have a significant national impact.

Reding and Shepard (8) have provided an estimate of potential energy savings in the six big fuel-consuming industries on the basis of short-term conservation measures (i.e., less than five years). Research and development on new processes, increasing product yields, new improved thermal insulations, and improved systems may yield additional conservation of energy, but the near-term estimates were made by assuming use of only presently available materials and techniques. The following discussion includes only what is now considered state-of-the-art and thus takes no credit for additional energy conservation that may accrue via research and development results on improved thermal insulations and applications of insulations. Reding and Shepard consider the order of effectiveness of conservation approaches in this immediate-term period to be (1) design modification, (2) insulation and maintenance, (3) process integration, (4) waste utilization, (5) process modification, (6) operation modification, and (7) market modification.

This discussion is focused primarily on item (2) and reviews the conservation potential in the six industries. The estimated immediate-term energy conservation potential in the six industries due to increased use of insulation and improved system maintenance is shown in Table 9.5. The estimated energy savings include those expected from steam system improvement plus a reduction of 10% in the fuel required for direct fired processes in each industry by increased use of insulation. The immediate-term conservation potential due to increased insulation and maintenance is obviously substantial, and the ratio of this value to the estimated total immediate-term conservable energy is conservatively of the order of a third of the total. The immediate-term conservable energy by this route is estimated to be about 5 to 10% of the process energy required by the six industries.

Table 9.5: Estimated Energy Conservation Potential in Immediate Period in Six Largest Energy-Consuming Industries

Industry	Estimated Energy Savings by Insulation . . and Maintenance*. . .		Estimated Total Industry Process . . . .Energy Usage**	
	10^{12} Btu/yr	10^{15} J/yr	10^{12} Btu/yr	10^{15} J/yr
Chemical	252	266	2,662	2,810
Primary metals	500	528	5,900	6,230
Petroleum	343	362	3,044	3,210
Paper	208	219	2,563	2,700
Structural clay products, glass, cement, others	190	201	2,000	2,110
Food	47	50	1,283	1,350

*Estimated precision, ±30%.
**Process energy only; does not include feedstock energy usage. Estimated precision, ±10%.

Source: TID-27120

Various recommendations for use of increased application of thermal insulation and maintenance in each industry in the immediate-term will now be discussed.

Immediate-Term Conservation Considerations

Increased application of insulation and improved insulation maintenance of steam systems in several of these industries have a significant energy conservation potential. These are indicated in Table 9.6.

Table 9.6: Estimated Immediate-Term Energy Conservation Potential in Energy-Intensive Industries*

Industry	. . . Estimated Energy Savings**. . .	
	10^{12} Btu/yr	10^{15} J/yr
Chemical	199	210
Primary metals	120	127
Petroleum	160	170
Paper	200	211
Structural clay products, glass, cement	40	42
Food	40	42

*By use of increased insulation and maintenance of steam distribution system.
**Estimated precision, ±30%.

Source: TID-27120

The estimated (6) reduction in chemical industry steam usage by this immediate-term fix would be about 20%. In addition to these estimated energy savings shown in Table 9.6, the industries could effect additional savings, indicated by comparing the estimated energy savings values for each industry in Tables 9.5 and 9.6. In the primary metals industries, use of the new oxide fiber insulation batting systems and more extensive use of skid rail insulations in metal heat treating furnaces could save considerable energy. More extensive use of thermal insulation in refineries, steam generators, steam strippers, reactors, flash columns, separators, and fractionators, which are typically poorly insulated, could save additional energy (6). The glass-cement-structural-clay-product industries can save additional energy by greater insulation of annealing and tempering furnaces as well as cement kilns (7).

Near-Term Conservation Consideration

Longer term (i.e., beyond 1980) energy conservation via use of thermal insulations could be increased in at least three ways. (1) Improved or more effective thermal insulations than those presently available would reduce heat flows for a given insulation thickness. (2) Improving the hot-wall refractories used in the aluminum, steel, and ceramic industries would allow lower effective furnace wall heat transfer, permitting higher inner wall temperatures without an adverse short-ening of the life of the dense refractories used in furnaces to contain molten metals and glasses or for densifying ceramics. (3) Castable and plastic mix refrac-tories for use in the steel, aluminum, chemical, and petrochemical industries having higher service temperatures, lower thermal conductivities, and /or im-proved mechanical properties would allow improved reactor vessel and furnace thermal insulation without significant shortening of the service life of the re-fractory linings.

It is obvious from the foregoing that the thermal insulation of directly fired process equipment and steam supply systems plays a major role in dictating energy flow in the energy-intensive industries. Improved insulations, more use of insulation, and more economical insulations could have a significant impact. Also, greatly improved insulation-weather barrier systems and installation practices could play a major role in upgrading current steam system performance.

OBSERVATIONS FROM INSULATION MANUFACTURERS, INSULATION USERS, AND EQUIPMENT AND SYSTEM DESIGNERS

Industrial Chemicals

One of the major insulation requirements is for steam lines, reactor vessels, and reactor piping. Temperature requirements for these applications typically range from ambient to approximately 250°C. Principal requirements of the insulation are (1) the insulation system must be essentially inert regarding flash point depres-sion, etc., for relevant organic liquids that may be spilled or leaked onto the in-sulation, (2) the insulation should have reproducible thermal transport properties (i.e., the properties should not vary significantly from batch to batch, etc.), (3) the insulation and its barrier system should be fairly resistant to mechanical abuse, (4) the barrier system should have a sufficiently low water leakage rate to prevent significant amounts of water from penetrating to the insulation, and (5) the insulation should be relatively easy to install on valves, because present systems for steam valve insulation are about three times as expensive as a similar length of straight pipe run. One large chemical firm is presently insulating every steam valve in its plants in spite of this cost.

One problem cited for this type of insulation, which at present is largely calcium silicate, is that the low mechanical strength of present non-asbestos-containing calcium silicate insulations leads to breakage, shrinkage, and hot spots. For example, friability of this material now often results in boxes of insulation arriving on the construction site in which all the pieces are broken. Another problem area is the inadequacy of the weather-barrier systems used with this type of insulation. Many users feel that none of the barriers now available are really watertight for the service life of the insulation. Also, installation practices used by installation contractors vary greatly.

There is apparently no certification by installation personnel and there is essentially no verification of the thermal performance of a system by the user, installer, or insulation vendor once it is installed. Both certification of personnel and verification of insulation performance were cited as very desirable by many insulation users. A few cases were cited in which process temperatures could not be maintained on a new system and additional heat tracing of lines had to be installed.

Several of the larger companies used computer programs to assist in thermal insulation designs. Critical factors in these design programs include constraints imposed by the individual corporation's accounting and financial system as well as specified factors related to current economic conditions in addition to heat transfer constraints.

Lack of knowledge on effects of currently available pipe insulations on ignition and detonation properties of various organic compounds important in chemical manufacturing was cited as a problem area. Also, there appears to be very little information available concerning degradation or other changes in pipe insulations with time after they are placed into service. Information on good insulation joint design, cements, and mastics is needed. Instrumentation of present systems for performance monitoring is essentially nil.

A new infrared imaging system (10) capable of detecting relatively small temperature and radiation differences on surfaces was mentioned as a possible way of identifying hot spots, open insulation joints, cracked insulation, etc. This equipment is presently quite expensive and has been evaluated by a few companies. More techniques such as this one or others will, however, be required in the future for routine monitoring of large systems if energy losses are to really be minimized.

Problems cited relevant to cryogenic systems were clearly separable into two categories. As discussed earlier, for very low-temperature cryogenic systems, the design and installation of these systems are typically contracted out to vendor specialists on a guaranteed performance basis. For intermediate low-temperature service applications, there is considerable discontent with expanded polyurethane because of its tendency to be permeated by water vapor, which subsequently deposits as ice at the cold surface. This progressively destroys the insulation properties of the urethane, although the warmer surface appears normal and dry. The vapor barriers available for low-temperature foamed urethane and polystyrene insulation systems are generally inadequate, and need to be greatly improved, and their performance needs to be verified in service. The toxicity of the vapors from burning urethane and its poor resistance to vertical flame spreading were cited as other reasons for general discontent with the urethane materials.

Foamed glass was cited as a superior insulation for use at intermediate low temperatures. Most of these low-temperature systems require a periodic heating and drying because of the inevitable moisture permeation and ice formation. The high-temperature stability of foamed glass relative to urethane or polystyrene foam is another significant advantage for the glass material. Foamed ceramics were mentioned as a new material that will be marketed soon as a high-temperature thermal insulation. It may be of some interest for these intermediate low-temperature applications as well. The available conductivity data for a foamed ceramic material at 205°C (400°F) are 0.14 W m^{-1} K^{-1} (1.0 Btu in. hr^{-1} ft^{-2} $°F^{-1}$), compared with 0.087 W m^{-1} K^{-1} (0.63 Btu in. hr^{-1} ft^{-2} $°F^{-1}$) for a foamed glass at the same temperature.

The foamed ceramic has a reported density about twice that of foamed glass. The crushing and flexural strengths of this foamed ceramic are very similar to those reported for foamed glass at a density of about 16 lb/ft^3 (260 kg/m^3). This material, with a maximum reported service temperature of 1150°C (2100°F) and low moisture permeation, may have some future applicability to low-temperature systems.

Primary Metals (Steel and Aluminum)

Thermal insulation problems in these industries appear to fall into two categories, (1) those associated with liquid-metal containment furnaces and metal reheat furnaces, and (2) those associated with steam and cryogenic systems. Steel companies use relatively large amounts of natural gas and electricity in addition to coal and steam. For improved refractory insulations for use in metal containment furnaces, it was apparent that for both steel and aluminum, reducing heat flow through a furnace wall by use of more or improved thermal insulating material on the outside may produce a serious shortening of the furnace hot-wall lining life as a result of this insulation reducing the heat flux through the wall and thereby increasing its temperature. The hot-wall lining life in a furnace is of major importance in steel melting and aluminum reduction and remelting.

Improved hot face refractories or changes in furnace design would be required before additional thermal insulation could be used effectively in these areas. Also, a great deal of the thermal losses associated with steel and aluminum are associated with the hot stack gases, (e.g., in one steel company, approximately 50% of the sensible heat lost in their processing is in the stack gases). It was felt that more attention should be paid to stack gas recuperation systems in addition to or in preference to thermal insulation. Another steel company is now assuming that energy costs will escalate at 4% per year for the foreseeable future and is factoring this into its thermal designs. Still another steel company is in the process of requesting bids on a 16" diameter (0.4 m) steam line a mile (1.6 km) long and for the first time requesting an insulation incremental cost analysis from the vendor.

Shortage of capital was cited as a serious detriment at this time to retrofit of insulation systems and construction of new systems by many companies. In most companies insulation projects should provide a return on capital in less than a year due to energy savings, but in many cases even these cannot be presently financed.

Furnaces associated with metal reheating are prime candidates for energy conservation via increased thermal insulation. The new insulating refractory fiber materials discussed previously are of great interest for these applications in the walls of these furnaces in the steel and aluminum industries.

Another example of a steel-heat-treating furnace application that results in considerable energy waste is the reheat furnace design in which steel slabs are conveyed through the furnace on water-cooled skid rails or skids. These rails, which are water-cooled pipes, are initially insulated on all but the top surface upon which the steel slabs rest. There is significant heat loss in the skid rails under this new condition, but historically the rail insulation tends to degrade rapidly with time and furnace heat losses to the rail system increase significantly.

Berg (11) has estimated that insulation currently available for this application lasts from four months to a year. It is often not repaired or replaced if it deteriorates or is damaged. Berg quotes a source who claims that about 50% of the water-cooled skid rails in heat-treating furnaces in the United States are fully insulated, while about 90% of those used in the major steel producing countries abroad are fully insulated. If insulation were applied to the presently uninsulated skid rail systems in the United States, the total saving of fuel would be equivalent to about 30,000 barrels of oil per day or 11×10^6 barrels of oil per year.

Another example of the energy savings potential of skid rail insulation systems is provided by Chart and Stock (12) in the following. It has been estimated that 22×10^6 Btu/hr (6.5 MW) will be lost to the cooling water in an average-size steel heat treating furnace with uninsulated rails. The cost of the insulation for a skid rail furnace is a function of the skid pipe size, but for a typical 4" diameter (0.1 m) pipe the cost of using insulating tile on the skids is about $ $24/ft ($79/m) for skid rails, $45/ft ($148/m) for crossovers, and $19/ft ($62/m) for verticals. In an average size furnace there is about 540 ft (165 m) of this water-cooled pipe, and an installation of this size would cost about $20,000.

Such an installation would save about 2.6×10^8 Btu per year per ft (9 J/m per year) of pipe or about $112,000/year for the entire 540 ft of pipe, based on a cost of $0.80/million Btu ($0.76/GJ). Thus, such a installation would pay for itself in about two months plus provide longer life to the skid rail pipes. This energy saving is about 1.4×10^{11} Btu/year (1.5×10^{14} J/year) or the equivalent of about 24,000 barrels of crude oil/year/furnace.

Petroleum Industry

Recognition of the economic value of and the corresponding need for effective insulation systems is currently high in the petroleum refining industry. This situation results from the awareness of fuel costs and energy cost trends and also from the application of energy flow analysis in unit operations used in refining plants. Other conditions contributing to the recognition of the value of thermal insulation are the significant cost of insulation systems applicable to petroleum industry applications and the growing awareness of problems and deficiencies in the performance of some insulation systems. It was noted that the cost of surface coatings, jackets, and insulation accounted for 4 to 8% of the capital cost of a new refinery plant.

There are several areas where insulation is being upgraded in existing plants. First, insulation thickness is being increased from 1 to 3 inches (25–88 mm). Second, insulation is being applied to valves, flanges, heat exchangers, and intermediate temperature storage tanks (fuel oil, waxes, residual oil, boiler feedwater) that have not been insulated in the past. Third, maintenance of insulation, especially protective coatings and jackets, is being stressed as the effects of moisture migration or insulation effectiveness and useful life are being realized. In all these areas, the actions are easily justified on a cost-effective basis.

Another attitude of interest is the preferential treatment of energy-conserving capital investments—improving thermal insulation, for example—compared with capital investments in general. Several companies required half the rate of return on investment for energy-conserving capital expenditures compared with other capital expenditures. This obviously gives a significant advantage to proposed insulation projects compared with other capital expenditures. Hence, energy-conserving process improvements are being implemented while other possible projects are postponed because of a general shortage of investment capital.

The large engineering staffs of the petroleum refining companies perform most of their insulation design and material selection; also product evaluation is performed by some companies as an input to material selection. However, there is essentially no verification testing of insulation thermal properties for installed systems; hence, vendor thermal property data for insulation materials are used directly without modification. ASTM standards are not considered adequate for design specifications and are supplemented heavily with company standards.

Insulation effectiveness is generally not verified in service, with a few exceptions in which cursory analyses were performed on newly insulated equipment. Some interest in doing more verification analysis is shown by some companies. Emissivity of stainless steel and aluminum jackets is thought to be higher in practice than indicated in published data.

Some problem areas that define requirements for improving the performance of various insulation materials have been identified. The major requirements are as follows:

(1) improved vibration resistance to breakage for calcium silicate insulations,

(2) improved resistance to slumping and compaction for fiber glass and mineral wools,

(3) improved moisture-resistant coatings and/or jackets for use in low- and medium-temperature applications.

Paper Industry

The paper industry is very positive toward the use of thermal insulation. There is an awareness of increasing fuel prices, but this situation thus far has not triggered much upgrading or economic reevaluation of insulation usage. This situation probably results from the much higher waste heat losses in exhaust gases and steam and water discharges than have been estimated for radiative and convective heat losses. Hence, the investment capital available for energy-conserving process modifications has been applied mostly in the area of waste heat recovery as opposed to thermal insulation.

Insulation system design and material specifications are contracted to architect-engineering (A-E) firms as a part of new plant design. Therefore, there is no significant use or criticism of ASTM insulation standards in this industry.

Architect-Engineering Firms

Since these firms provide engineering design and consulting services to the manufacturing and electric utility industries, their main interest and expertise is in thermal and mechanical design of insulation systems as opposed to operational and service-life considerations. These firms typically have two types of contracts. One form of contract is fixed price, which often involves offering a customer a relatively standardized design. The other form of contract is the cost-plus-fixed-fee type.

The fixed-price type often tends to result in lower initial capital costs to the customer due to savings incurred in previous design work and construction experience. The cost-plus-fixed-fee type, which would be involved in specific insulation optimization, is typically more costly. Many firms prefer to minimize capital costs of new construction even if it results in future higher maintenance costs during the plant or system life because of tax statutes, etc. This situation tends to cause A-E firms to offer proposals involving minimum capital costs and often involving designs that are not particularly energy conservative. Therefore, the A-Es economic interest would be to minimize the capital cost of insulation systems within the economic design criteria imposed by the plant owner. Operational and maintenance considerations would tend to be secondary to capital cost considerations in design and material selection.

An important design requirement in selected industries is imposed by the OSHA standard that requires steam and hot water lines to be insulated or guarded in situations where personnel contact is possible. The choice of jacket or coating material placed over the insulation is also important in designing the insulation to limit surface temperature. ASTM standards are often used in material specifications to provide a common basis or background for different specifications. However, additional specifications prepared internally are added especially for installation and inspection procedures.

One situation of a new insulation requirement was stated as follows. In nuclear power plants, motion-limiting supports are placed around the primary system piping to confine the discharge flow area of a hypothetical circumferential pipe rupture. The thermal insulation material placed between the pipe and the support needs to be a crushable material to allow for normal differential movement between these elements.

Cement Industry

As indicated earlier, cement manufacture requires significant use of energy in the United States. It requires about 7.2×10^6 Btu to produce a ton of cement

(6.9 MJ to produce a kilogram). Europeans have reduced fuel consumptions to much lower levels than the general case in the United States by using improved techniques of heat transfer in cement kilns. Extensive use of heat recuperators was the principal method by which Europeans have reduced the energy required to produce cement. A large kiln of this type in Europe has demonstrated the capability of producing cement with 3.3×10^6 Btu/ton (3.2 MJ/kg). This is only about 46% of the energy required in the United States. Fluidized-bed processing in cement manufacture also offers some additional gains in fuel efficiency, but this process technique has apparently not been evaluated very extensively in the United States. As much as 20% of cement production costs are often represented by fuel costs, and thus this industry will be relatively sensitive to potential cost savings possible by conserving fuel. Increased usage of thermal insulation in this industry would be primarily in improved kiln linings and steam supply systems.

Glass Industry

The glass industry consists of three major segments. These are (1) glass containers, (2) flat glass, and (3) pressed and blown glass. Because of the nature of the flat glass and the pressed and blown glass segments of the industry, it is very difficult to obtain energy balance data because this information is usually considered to be proprietary. Battelle Columbus Laboratories (3) estimates that for 1973 the flat glass segment of the industry used about 67.2×10^{12} Btu (71.0×10^{15} J) and that the glass container industry used 21.6×10^{12} Btu (22.8×10^{15} J).

Most of the flat glass manufacture in the United States has converted to the Pilkington float glass process, in which molten glass leaves the furnace on a bed of molten tin. This process has superior production economics and production capacity compared with the previous flat glass process in which the plate was formed by passing glass in the viscous stage through metal rolls as it left the furnace and was subsequently polished on both sides. This process change has resulted in significant energy savings because of higher production yield. All the furnaces used in glass manufacture except for the glass melting furnace suffer the large stack heat losses common to other industrial furnaces.

The glass melting furnaces are typically large and employ regenerative heat recovery stages, which are often much larger than the furnace tank itself. The result is that large glass melting furnaces may be the most efficient of present large high-temperature industrial furnaces. A possible rival is the basic oxygen steel furnace.

The glass industry is concerned with energy losses from steam systems as well as annealing and tempering furnaces. Some of the new fibrous insulations and more extensive use of recuperators could result in significant energy savings in this industrial sector. Additional thermal insulation of large glass-melting furnaces is not very practical since they make extensive use of recuperators and additional furnace wall insulation would raise the refractory hot-face temperature and result in more rapid degradation of the refractories that contain the molten glass.

Food Industry

Since the food industry requires of the order of 10^{15} Btu/year (10^{18} J/year) it should be examined for the potential of energy conservation. From Table 9.5 the estimated potential energy savings in the food industry is approximately 47×10^{12} Btu/year (50×10^{15} J/year), and almost all of this is heat available at 250°C or less. This energy loss is essentially all from steam lines and food process lines. Very few steam or process lines are insulated in the food industry because of stringent cleanliness requirements.

Large food producers indicated that they typically use bare lines because of the requirement for frequently washing down and sterilizing a facility. They are not aware of any thermal insulation system that would be impermeable to moisture from frequent washdowns, contribute no particulate matter that might contaminate food, harbor no bacteria, and be flexible under certain piping design requirements.

Usually, no dead-end pipe runs are permitted where food products are involved, and this requires process pipes to be readily disconnected and reconnected to other points in the process stream. Development and demonstration of such an insulation system would be welcomed by the food industry and could result in appreciable energy savings.

Electric Power Utilities

There is a general awareness of the value of thermal insulation in electric power utilities by virtue of the fact that "energy" is their product. However, there is quite a range in the desire to improve thermal insulation performance. Actions taken to decrease heat losses include (1) increasing insulation thickness on valves and flanges and (2) improving fit of insulation sections.

Design procedures vary from subcontracting the entire thermal insulation design based on calculated average heat loss to performing design with in-house engineering personnel. Design of reflective metal insulation for nuclear plants is all subcontracted presently. There is a general feeling that thermal property data are adequate for design, fostered by the subcontracting method of obtaining insulation design. ASTM standards are found to be too general and so are supplemented extensively with company specifications.

Performance verification is not done extensively. Performance is verified in some cases on fossil-fueled plants and on nuclear plants before criticality with system heat balance and heat rate meters. The tests show that hot spots do occur with metallic reflective insulation and calcium silicate insulation at high temperatures.

Insulating odd shapes effectively can be difficult and expensive. The turbine flange was mentioned as one such difficult area to insulate.

SUMMARY OF TECHNICAL PROBLEMS

Most of the problems identified in this assessment pertain to specific industries and manufacturing processes. In general the technical problem areas where some near-term energy savings could be accomplished are:

(1) improved steam pipe insulations of the calcium silicate or competitive types with lower thermal conductivities and improved fracture and dusting properties relative to present products,

(2) greatly improved pipe insulation weather barrier systems to prevent water penetration and degradation of the thermal conductivity,

(3) improved and standardized pipe insulation installation practices,

(4) cryogenic insulation systems where all pathways of condensible vapors to the cold surface can be closed with certainty during the operating life of the system,

(5) greatly improved pipe insulation techniques for steam system valves and flanges,

(6) greatly improved steel furnace skid rail thermal insulations,

(7) better data on the emittance properties of high-temperature thermal insulations and weather barrier materials,

(8) data on changes that occur to the total emissivity of high-temperature insulations during service life,

(9) data on changes that occur to the thermal conductivity of high-temperature insulations during service.

Another problem area involves thermal insulations for use in liquefied natural gas (LNG) land-based storage tanks and seagoing LNG tankers. Several of these installations have been constructed with polyurethane foam used as the thermal insulation. Moisture permeation and subsequent loss of insulation quality due to ice formation have been observed. Two instances were identified in which two tankers were being built and insulated, one with balsa wood and the other with foamed polystyrene rather than polyurethane. The combustion potential of large tankers and storage vessels when insulated with polyurethane foam was cited as an item of concern by several contacts. One large LNG distributor has banned polyurethane-insulated tankers from their terminals.

From the foregoing, it is clear that the thermal conductance properties of available thermal insulations must be better understood as a function of service conditions and service life. For the case of pipe and tank insulations, the performance must be regarded from a total system point of view, including the environmental barrier system, mastics, joint design for the insulation, installation practice, and the thermal transfer properties of the insulation material. A more quantitative understanding of the role of each of these variables on insulation system performance will be required before systematic and predictable improvements in energy conservation in the low and intermediate temperature ranges can be effective.

For the case of high-temperature applications, the insulation must also be regarded from a system point of view. The thermal performance of aluminosilicate and single-oxide fibrous insulations should be quantitatively demonstrated in various furnace wall designs for selected environments as soon as possible and the results of these performance tests given wide distribution. Similarly, insulating firebricks should be examined as to property and performance variability and a critical study made to determine the possibility of significantly reducing the thermal transport for various classes of IFB. Current furnace wall designs should be examined to establish if heat transfer through the walls can be better utilized than in the current practice of allowing the heat to flow through to the cold face and be lost. This might be accomplished via heat recuperator channels within the walls. Insulation systems for aluminum reduction and remelting, iron reduction, and steel reheating should be critically examined and improved via furnace design improvements and insulation improvements, followed by quantitative field demonstrations of these improvements.

Field demonstrations in these areas have typically been accomplished outside the Federal sector. They have typically resulted from agreements between an interested potential insulation user and an insulation or refractory vendor. Because of cost and time constraints, the user is forced to rely heavily on vendor-supplied data and by necessity usually chooses one insulation and/or refractory for actual demonstration. Results from such demonstrations may be made known or restricted depending on the value of the results to the user-demonstrator. This process results in a relatively slow dissemination of such information.

Selected field demonstrations supported entirely or in part with Federal funds with subsequent rapid dissemination of the quantitative results to industry could be one technique for relatively rapid performance comparisons of thermal insulations and thermal insulation systems.

CONCLUSIONS

(1) The quantitative performance of low-, ambient-, medium-, and high-temperature thermal insulation systems is not adequately known as a function of the materials used, the design parameters, the installation practices, or the age of the system.

(2) Thermal and mechanical properties of insulation materials such as calcium silicate are deficient.

(3) The combustion risks associated with organic foam insulations have not been completely evaluated.

(4) The thermal transport properties of inorganic fiber insulations currently on the market are not well understood or optimized for performance.

(5) Certification of thermal insulation installers has not been extensively used as a means of standardizing installation practices.

(6) The insulation manufacturing industry currently feels that its products are either totally suitable for current needs or that it has new products developed that will satisfy most needs when market conditions develop sufficiently.

(7) New total insulation system performance testing techniques such as infrared scanners have not been extensively evaluated, developed, and utilized.

(8) Contacts in the steel industry felt that more attention should be concentrated on stack gas heat recuperators as well as on thermal insulations.

(9)   Verification testing of the properties of available industrial thermal insulations would be very helpful to users and designers. ASTM recommended standards and specifications were not considered adequate by many contacts.

(10) The types of contracts used to employ architectural engineering firms and the minimum capital cost incentives imposed by industrial customers of A-E firms tend to develop plant and system designs that are not optimized for minimum heat losses via use of thermal insulations.

(11) Easily sterilized reflective types of thermal insulation have not been considered for possible use in the food industry.

(12) Immediate industrial energy conservation potential due to increased insulation and improved maintenance is estimated to be of the order of 5 to 10% of the present industrial process energy consumption.

REFERENCES

(1) O'Keefe, W., "Thermal Insulation," *Power* 118(8): 21–44 (August 1974).

(2) Myers, J.G., et al., *Energy Consumption in Manufacturing* (A Report to the Energy Policy Project of the Ford Foundation), Ballinger Publishing Company, Cambridge, Mass., (1974).

(3) Battelle Columbus Laboratories (Report to U.S. Bureau of Mines), DB-245579 AS, *Energy Use Patterns in Metallurgical and Nonmetallic Mineral Processing (Phase 4 — Energy Data and Flowsheets, High Priority Commodities)* (June 27, 1975).

(4) Battelle Columbus Laboratories, (Report to U.S. Bureau of Mines), DB-246357 AS, *Energy Use Patterns in Metallurgical and Nonmetallic Mineral Processing (Phase 5 — Energy Data and Flowsheets, Intermediate-Priority Commodities)* (Sept. 16, 1975).

(5) Reding, J.T. and Shepard, B.P., *Energy Consumption: The Primary Metals and Petroleum Industries,* Dow Chemical Company, PB-241 990 EPA-650/2-75-032-b (April 1975).

(6) Reding, J.T. and Shepard, B.P., *Energy Consumption: The Chemical Industry,* Dow Chemical Company, EPA-650/2-75-032-a (April 1975).

(7) Reding, J.T. and Shepard, B.P., *Energy Consumption: Paper, Stone, Clay, Glass, Concrete and Food Industries,* Dow Chemical Company, EPA-650/2-75-032-c (April 1975).

(8) Reding, J.T. and Shepard, B.P., *Energy Consumption: Fuel Utilization and Conservation in Industry,* Dow Chemical Company, EPA-650/2-75-032-d (August 1975).

(9) Shell Oil Company, *The National Energy Outlook,* 9th Printing, (January 1974).

(10) Eklund, J.K. and Baeu, D., "An Infrared Eye," *Ind. Res.* 17(4): 52-55 (April 1975).

(11) Berg, C.A., "Conservation in Industry," *Sci.* 184(4134): 264-70 (1974).

(12) Chart, A.L. and Stock, D.F., "Cost Factors Related to Energy-Saving Applications of Refractories," Harbison-Walker Refractories, Pittsburgh, Pa. (unpublished manuscript, October 1975).

ECONOMIC CONSIDERATIONS
OF THERMAL INSULATION

The material in this chapter was excerpted from reports prepared by York Research Corporation (PB-259 937) and by Oak Ridge National Laboratory (TID-27120).

INTRODUCTION

This chapter presents a brief review and summary of economic considerations and analytical procedures that currently exist in the thermal insulation field. Many other economically significant factors can also be considered; among these are labor costs for installing insulation systems, manufacturing cost variables, and health considerations in using insulation materials.

BACKGROUND FACTORS AFFECTING INSULATION USE

One of the most dramatic changes in recent years is in fuel price increases that have occurred since the early 1970s. Such rapid price increases have improved the economic value of fuel conserving practices such as the use of thermal insulation. However, in making analyses of the economic value of insulation over lifetimes of 10 to 15 years, assumed fuel escalation rates become very speculative. Therefore, the sensitivity of economic insulation thickness to projected fuel prices is of interest to companies as they plan for future expansion of production facilities.

Another factor that can affect the use of thermal insulation is the availability of a traditional industrial fuel, such as natural gas. With the supply of natural gas for industry becoming more questionable in times of high domestic demand, the efficient use of insulation can become an important factor in retaining production continuity for industries dependent on natural gas, such as steel and industrial chemicals. As a result of the Arab oil embargo and the emphasis on national independence of energy supplies, energy conservation in industrial processes has received much attention. The potential for reducing the waste energy is being realized in most industries; also, national goals for energy conservation are stimulating the use of more efficient processes. Hence, the application of thermal insulation to economically justified thicknesses is one measure that industrial companies can take to improve conservation of energy in the immediate time frame.

ASSESSMENT OF ECONOMIC INSULATION THICKNESS ANALYSES

The value of insulation in industrial processes is derived from several different economic benefits. Direct economic benefits or savings result from conservation of fuel, conservation of products, and increased production of products. Indirect economic benefits are obtained from better control of temperature and protection of personnel from hot surfaces.

The economic value of industrial insulation from the direct benefits appears to be so obvious that it has been stated that the question is not whether to insulate but how much insulation is justified. This section, then, briefly reviews: (1) the traditional procedure used to determine what insulation thickness is economically justified—the "economic thickness"; (2) the application of economic thickness analyses by the Thermal Insulation Manufacturers Association (TIMA) and by industrial companies; and (3) current economic conditions and considerations that were found to be important to industrial users of insulation.

Traditional Analysis of Economic Thickness

One of the first analytical procedures for calculating the economic thickness of insulation was presented by L.B. McMillan in 1926 at a meeting of the ASME. The detailed development of McMillan's economic thickness analysis is readily available in Malloy's book (1) and hence is not reproduced here. However, some of the more important features of this analysis procedure are presented before reviewing the present status of economic evaluations of industrial users and TIMA.

The basic cost factors incorporated in the traditional analysis of economic thickness of insulation are grouped into two types. One type of cost factor decreases with additional insulation thickness, reflecting cost savings from additional insulation use. Such decreasing cost factors for a new plant are:

> cost of fuel and/or purchased power,
> capital investment in heat (or cold) producing and distributing
>   equipment,
> value of money for capital investment,
> interest on investment, and
> marginal maintenance cost for plant equipment.

The cost factors that increase with additional insulation thickness are similar to the decreasing factors, as follows:

> capital investment for installed insulation,
> value of money for investment,
> interest on investment, and
> marginal maintenance cost for insulation.

To compare the increasing and decreasing costs for a specific insulation config-
uration, all cost factors are put on an annual average basis using assumed lifetimes or depreciation periods for the plant and insulation investments. Appropriate average values of fuel and maintenance costs are chosen by the analyst
and may reflect escalation of such costs over the project life.

Another cost factor chosen at the discretion of the analyst is the capital investment money value, expressed as a percent per year of the capital investment.
This factor represents a value of capital to a company on the basis of the potential investment options available to the company. (Similar economic terms for
the value of money for invested capital are the "required rate-of-return" and the
"discount rate".)

The criterion for selecting the economic thickness of insulation is based on the
simple concept of determining the first ½" (13 mm) thickness increment for
which cost increase equals or exceeds cost decrease. Referring to Figure 10.1,
the economic thickness occurs at the minimum total annual cost since the lost
heat cost and insulation cost are monotonically changing cost functions.
McMillan's economic thickness derivation was therefore based on analytically
determining the insulation thickness at which the minimum total annual cost
occurs.

Figure 10.1: Economic Thickness of Insulation

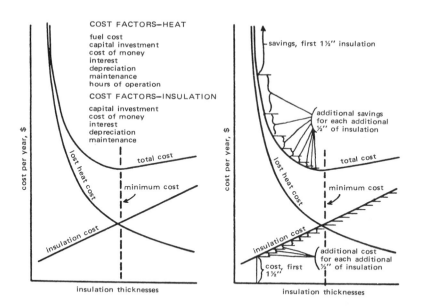

Source: TID-27120

Cost Savings from Insulation Capital Investment

In addition to the minimum total cost criterion of McMillan's analysis, industrial companies often include the requirement that the reduction in lost heat

cost be a certain percentage per year of the capital invested in insulation over the life of the insulation. This criterion is called the required return-on-investment (ROI) and can be applied to either the entire insulation thickness or the last ½" (13 mm) increment.

In the former case, the ROI is an overall value representing the total savings from an uninsulated condition; for the latter case, the ROI is an incremental value for the last ½" of additional thickness. By requiring a minimum incremental ROI, a company may be assured that each additional increment of insulation can produce savings that are at least competitive with alternate investments. However, the minimum incremental ROI can be set so high that the economic thickness, in terms of minimum total cost, is not reached. Then the minimum required ROI becomes the criterion that determines the maximum insulation thickness justified economically.

From Figure 10.1, one can see that the incremental savings decrease and the insulation investment increases with additional insulation thickness. For a new plant, the overall ROI for insulation typically decreases from several hundred percent per year for the first increment of insulation thickness to about 100% per year at the economic thickness.

Such ROI values result largely from the fact that, except for small pipes, total invested capital for insulation and process heat generating capacity decreases with the use of insulation at the economic thickness relative to the uninsulated condition (1). Even with no savings credited for the reduction in steam generating capacity investment, overall ROI values (from Malloy) for ½" (13 mm) pipe insulated to the economic thickness ranged from 19% per year at 100°C (212°F) to greater than 200% per year above 480°C (900°F).

In addition to the attractive ROI values obtained for the overall economic thickness of insulation, incremental ROI values were reported (1) to be at least 19% per year for the last ½" thickness increment at the economic thickness. The ROI values reported by Malloy are from studies performed with early 1960 economic cost conditions. Although no additional economic calculations were performed as a part of this assessment, it is believed that the current ROI potential for industrial insulations is comparable to, if not better than, the ROI for early 1960 conditions.

Discounted Cash-Flow Investment Analysis

The discounted cash-flow (DCF) method is a more sophisticated method being used by some industrial companies to analyze the profitability of investments in various situations. Cash flow refers to the timing of disbursements and receipts for capital and operating costs and earnings or savings during the assumed lifetime of the project. The main advantage of the DCF method over the average annual cost method employed in McMillan's original analysis is that the effect of timing of capital expenditure on resulting earnings is accounted for.

All the cost factors used in McMillan's analysis are also used in the DCF method. The value of money is represented by an assumed "discount rate", which is applied to all cash flows to put them on the same "present value" basis. When all present value cash flows are summed over an assumed project life, the "net present value" (NPV) of the project investment is obtained. The prime objective of this method is to determine a discount rate for which the NPV equals zero at the end of the assumed project life. The discount rate for this condition is called the discounted rate-of-return (DRR). Stated another way, when the DRR is equal to an assumed discount rate, then the risk is zero for not recovering the invested capital during the project lifetime. [See Bierman and Smidt (2) for more detailed discussion of the DCF methodology.]

The DCF method is discussed because the DRR is used in a manner similar to the incremental ROI to determine how much insulation is economically justified. Companies employing the DCF method also use it to compare the profitability of different capital investments.

YORK RESEARCH CORP. ANALYSIS OF ECONOMIC INSULATION THICKNESS

This report, PB-259 937, has calculation procedures that are probably the most useful for analysis of economic insulation thickness by industrial users. First, the discounted cash-flow method is used to obtain the economic thickness and net present value basis. Second, insulation cost factors input by the user allow for single, double, and triple layers and also for three complexities of pipe fittings. Third, cost of money and other cost factors can be varied independently over a wide range. This section briefly describes the analysis and computational procedures as presented by York Research Corp.

Economic Analysis for a One-Material System

The primary function of insulation is to reduce the loss of energy from a surface operating at a temperature other than ambient. The economic use of insulation reduces plant operating expenditures for fuel, power, etc.; improves process efficiency; and increases system output capacity or may reduce the required capital cost.

In determining the most economic design for an insulation system, two or more insulating materials may be evaluated for least cost for a given thermal performance; or, optimum insulation thickness may be selected for a specific insulation type. In either case, the decision should be based on which design will save the greatest number of dollars over a specified period, in both initial and continuing costs. There are two costs associated with the insulation type chosen. For any given thickness, there is:

> a cost for the insulation itself; and
> a cost for the energy lost through this thickness.

The total cost for a given period is the sum of both costs. The optimum economic thickness is that which provides the most cost effective solution for insulating and is determined when total costs are a minimum. Since the solution calls for the sum of the lost energy and insulation investment costs, both costs must be compared in similar terms. Either the cost of insulation must be estimated for each year and compared to the average annual cost of lost energy over the expected life of the insulation, or the cost of the expected energy loss each year must be expressed in present dollars and compared with the total cost of the insulation investment. The former method, making an annual estimate of the insulation cost and comparing it to the average expected annual cost of lost energy, is the method used here.

Cost of Lost Energy

The rate of energy transfer through the insulation, the cost of value affixed to that energy, and the operational hours per year determine the cost of lost energy per annum. The rate of the energy transfer is a function of the following:

> the temperature difference across the insulation,
> the thermal conductivity,
> the thickness, and
> the thermal resistance of the external surface of the insulation.

The value of energy in a system is directly related to the end use of that energy. For example, high-energy steam capable of driving a turbo-generator has a greater economic value per Btu to a utility than low-energy steam, which cannot be utilized to produce electricity. Another method for determining the value of energy is to add the costs of producing the energy, which is the method used here. The cost of purchasing or producing energy includes fuel or power costs, capital equipment costs, and operating and maintenance expenditures.

Fuel or power costs have the greatest effect upon the dollar value of energy. The efficiency with which these fuels are converted into process heat affects the value assigned to energy in a process. The energy conversion efficiency or co-efficient of performance (refrigeration process) of the equipment must be considered. Since fuel and power costs will most likely change with time, the average cost of each over the life of the insulation project is used rather than today's costs.

The energy being conserved because of insulation requires capital equipment for its production. The capital cost of the energy plant must, therefore, be assigned to the dollar cost of energy. This cost is estimated per annum considering the plant depreciation period, the average annual energy production and the cost of money (the cost of financing the investment). Annual maintenance and operating expenses also contribute to the cost of energy, and must be included.

Cost of Insulation

The cost of insulation is the sum of the annual insulation investment cost and the yearly insulation maintenance expense, including material prices and labor for installation. This is also valid for installation of retrofitted insulation. The initial cost of the insulation system is prorated over the chosen project life using the appropriate cost of money or required rate of return on the last increment of insulation applied.

The period over which the insulation investment cost is considered is a factor in selecting the economic thickness. If the chosen period is short, the annual insulation cost will be high, the economic thickness will be small, and the insulation system will not provide the lowest total annual cost over the service life of the insulation. It is therefore recommended that the insulation service life be used for the project period if user project payout periods do not otherwise dictate.

It should also be noted that in addition to economic considerations, the thickness that is calculated for pipes and equipment that operate at subambient temperatures must be sufficient to prevent condensation. Minimum thickness to prevent condensation is based upon total heat gain to the insulation surface from radiation and convection at the dew point temperature corresponding to design ambient air conditions.

Theory

Optimal economic insulation thickness may be arrived at by two methods: the minimum total cost method and the incremental (or marginal) cost method. For a given situation both methods will yield the same thickness solution.

The minimum total cost method involves the actual calculation of lost energy and insulation costs for each insulation thickness. The thickness producing the lowest total cost is the optimal economic solution. However, the numerous calculations involved in this method require a computer for derivation of an answer; this precludes the use of the minimum total cost method in this description.

The incremental or marginal cost method provides a simplified and direct solution for the least cost thickness without having to calculate total annual costs. With this method, the optimum thickness is determined to be the point where the last dollar invested in insulation results in exactly $1.00 in energy cost savings, on a discounted cash flow basis.

The "incremental cost" is a term applied to the change in installed cost between two successive insulation thicknesses. At thickness L, the cost for adding the insulation thickness (ΔL), is given as $m_c = \Delta C/\Delta L$ (see Figure 10.2 and compare with Figure 10.1) where m_c is incremental insulation cost ($/in); ΔC is difference in insulation installed cost ($) for thicknesses L' and L"; and ΔL is additional insulation thickness, L" − L' (in). The dollar investment required to increase the insulation thickness from L' to L" is therefore determined as m_c.

At thickness L, the reduction in the cost of lost energy obtained by adding the insulation thickness ΔL, is given as $m_s = \Delta S/\Delta L$, where m_s is incremental savings in the cost of lost energy ($/in); and ΔS is difference in the cost of lost energy ($) between thicknesses L' and L". The dollar savings in energy obtained by increasing the insulation thickness from L' to L" is therefore determined as m_s, which always has a negative value because adding insulation reduces lost energy cost.

The total annual cost is equal to the sum of the cost of lost energy and the cost of insulation. The change in total costs when additional insulation is added is equal to the sum of m_s and m_c. When total cost is minimum, the change in the total cost is equal to zero and m_s equals m_c. When this condition is met, adding more insulation is not profitable because the incremental cost for additional insulation, m_c, becomes greater than the dollar value of the energy saved, m_s, by that additional insulation.

Figure 10.2 depicts both the minimum total cost and incremental methods and reflects the requirement for multiple layers needed in higher insulation thicknesses and/or needed to alleviate expansion and contraction forces in cyclical temperature systems.

Figure 10.2:   Insulation Thickness vs Insulation Cost

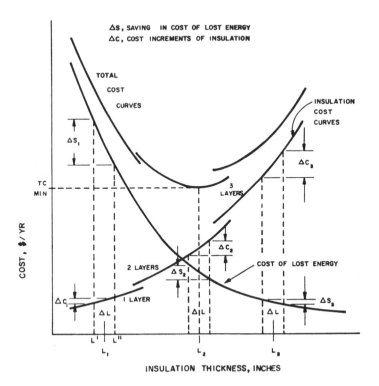

INSULATION THICKNESS, INCHES

Source: PB-259 937

Figure 10.2 also depicts typical insulation cost curves for single, double, and triple layers of thickness. As shown, increments of insulation thickness cost more as total thickness increases either within a single layer or as a result of multiple layering. The average slope of the triple layer cost curve is greater than the average slope of the double layer cost curve. The reason is that both material and labor costs increase with thickness and with each additional layer. The step between the respective insulation cost curves is primarily the labor cost to add the additional layer.

The total cost curves are for single, double, and triple layers, and are generated by adding the insulation and lost energy cost curves. For the example shown in Figure 10-2, the optimal economic thickness falls in the double layer range, as total costs are a minimum at point L_2. The insulation thickness, L_2, is the economic solution using the minimum total cost method.

The incremental method is demonstrated in Figure 10.2 by comparing the change in insulation cost (ΔC) with the savings in cost of lost energy (ΔS) resulting from the addition of an insulation thickness (ΔL). At thickness L_1, the addition of ΔL causes a large saving in lost energy (ΔS_1) at a small increase in insulation cost (ΔC_1). At thickness L_3, the addition of ΔL is rather expensive (ΔC_3) and saves only a small amount in energy costs (ΔS_3). The optimal economic thickness is arrived at when the last dollar invested for insulation results in one additional dollar in energy cost savings. This condition is met at thickness L_2, where ΔS_2 and ΔC_2 resulting from the addition of ΔL are equal, ($m_c = m_s$).

Life Cycle Costing

A popular concept of accounting for costs is the "life cycle" cost approach where the stream of outlays is added for the expected life of the project and the total cost over that life is compared with other alternatives. The cost analysis of this description can be used in this mode as shown below.

The insulation project considers the cost of thermal losses and the cost of insulation together to solve for least cost insulation thickness. The thermal losses are the product of the heat flow and value of that heat through the insulation. The cost of insulation considers the material, labor, and maintenance for the installation. The thermal losses are given as heat flow per unit of insulation; the unit is either a linear foot of pipe or a square foot of flat surface. The heat flow through the insulation is assumed to be constant over the life of the installation. The cost of the heat is taken from the price of heat or fuel, conversion efficiency of fuel, heating value of fuel, and the fact that prices of heat or fuel are expected to change over the years of the project life.

In life cycle costing the procedure would be to take each year of life and escalate the cost of the heat for each year and then sum those costs. Here, a multiplying factor, A, is provided, which sums all of the annual compounding multipliers and divides by the number of years of life to give an average multiplier for the term of the project. This saves calculating the stream of costs, adding them, p, and dividing by the number of entries to get an average. Thus, this presentation does the life cycle cost analysis for the cost of heat.

The cost of installed insulation must also be reduced to an average annual cost in order to compare it with the heat cost. Since the cost of installed insulation is an initial lump sum, that cost is multiplied by a factor, B_3, which gives the annual equal payment that will just return the capital and interest at the end of the project life. This uniform payment principle is also used in the life cycle costing for plant capital expense. For the sake of simplicity, the insulation maintenance costs are estimated to be 10% of the annual installed insulation cost for each year and added in as such.

It is evident then that this presentation provides a least cost insulation thickness based upon the principles of life cycle costing (see Figure 10.2). To achieve this, the compounding interest rates for the factor A (to adjust heat cost) and for the factor B_3 (to amortize the capital investment of the insulation cost) must be equal to the rates used in the life cycle analysis. (B_2, used to amortize the cost of a heat-producing facility, is calculated using an interest rate, i_2, and term of the facility, n_2, both of which may be different from the rate and term used to calculate B_3.)

Present Value (PV) Analysis

For the cost analyst who prefers to use present value as the proper yardstick for evaluating projects, the method presented here does just that, although it may not give the first impression of doing so.

Present value brings future cash flows back to the present by discounting at desired rates. The solution provided here gives an identical answer by forcing the insulation thickness chosen to provide the least expensive solution for the sum of heat costs raised to inflated levels and the installed insulation costs raised to average cash levels using the cash of the future.

To match a true present value analysis it is only required that the B_3 multiplication factor use the life term in years that the PV analysis uses and that the i_3 of the B_3 factor match the discount rate of the PV solution. The discount rate can be that for mere inflation or for an opportunity cost of money or for a required return on investment. The last increment of the calculated economic thickness will return that chosen rate.

COMPUTER ANALYSIS PROGRAMS

The economic thickness solution developed by McMillan requires an iterative solution involving the total monetary value of heat, the cost of insulation per unit of thickness, the physical dimensions of the system, the external thermal resistance, insulation thermal conductivity, and the operating and ambient temperatures. With so many variables involved, the use of computers to calculate the economic thickness was an essential step in applying McMillan's analysis to the wide range of real world insulation conditions.

A joint computer calculational effort between Union Carbide Corporation and the West Virginia University produced a manual (3) by the National Insulation Manufacturers Association (NIMA) in 1961. This calculational program was greatly simplified by the observation that insulation cost at a specific time and location increases linearly with thickness, with a constant value for the intercept at zero thickness.

Subsequent to the NIMA publication of 1961, an updated economic thickness manual (4) called ECON-I was published in 1973 by the Thermal Insulation Manufacturers Association (TIMA). It followed the same calculational procedure that was used in the NIMA manual with the following differences.

The NIMA manual required the user to determine the average annual heat and maintenance costs on whatever basis he chose. However, ECON-I internalized a 7.5% per year interest rate on an annual insulation maintenance cost of 1% of the total plant investment. A heat cost escalation rate of 4% per year was also included in ECON-I with the capability to adjust for escalation rates up to 10% per year. Base insulation cost factors were also updated to 1973 conditions in ECON-I.

In 1975, TIMA published (5) an economic thickness manual called R-ECON for adding or retrofitting insulation thickness to the economic value projected at that time. It followed the same calculational procedure as ECON-I with new values assumed for the following variables: 7% per year escalation rate on heat cost; 10% per year interest rate; $20.00 per lb/hr ($44.00 per kg/hr) of steam-generating capacity. In addition, the range of heat cost was limited to $2.20 per 1,000 lb ($4.84 per metric ton) of steam ±20%, and the selection of insulation cost was limited to three values: a base cost, base cost + 25%, and base cost + 50%. Therefore, R-ECON updated some of the economic data to more current values, but at the same time restricted the range of several important cost factors.   Contacts with large industrial users of insulation and also engineering consulting firms noted little or no use being made of the R-ECON and ECON-I manuals for economic thickness. The reasons for this lack of use varied greatly. The most important reasons are listed below.

> Confidence was lacking in the procedure for adjusting insulation costs for different thicknesses and specific installation conditions. Many companies keep insulation cost data for their locality and installation conditions so that they can more accurately determine economic thickness with their company computer programs.
> Flexibility was lacking in investigating cost of money factors. Many companies wish to explore the sensitivity of capital investment projects to variations in the cost of money.
> Discounted rate-of-return and net present value are used for company investment decisions. Companies using this method will analyze investment in insulation in the same way other investments are analyzed.
> Economic thickness specified by ECON-I or R-ECON can be impractical to apply in some cases. This situation results in not using the recommended results rather than not using the procedure. The nature of the problem is that such large thicknesses are specified that confinement occurs between pipes in pipe runs and between nozzles on tanks.

SURVEY OF ECONOMIC CRITERIA OF INDUSTRIAL USERS

Many industrial companies have discussed several economic criteria that apply to current investment decisions. For companies using discounted cash-flow analysis, the discounted rate-of-return (DRR) required for new capital projects traditionally ranged from 10 to 15% per year. However, with increasing competition for capital, some companies have increased their required DRR to 30 to 50% per year.

Even with such high earnings requirements, insulation investment has been justified for the economic thickness determined on a minimum total cost basis. In some petroleum refining companies, the required DRR for energy conserving investments such as insulation have been reduced to about 15% per year. Again, investment in thermal insulation usually meets such a requirement.

The payback period is a traditional measure of the time required to recoup invested capital. Where payback periods have traditionally been required to be 5 to 6 years, in many instances the required payback period has been reduced to 1 to 2 years. Many additions of insulation thickness cannot meet such a short period requirement. However, insulation of bare valves, flanges and tanks usually can satisfy a 1 to 2 year payback period requirement.

The final and basic economic criterion that current investment projects must satisfy is simple availability of the capital to proceed with a project. Even though an investment may be judged desirable on a profitability and short payback basis, in some cases, the lack of available capital at acceptable interest rates has postponed further use of insulation during current economic conditions.

OBSERVATIONS AND CONCLUSIONS

As a result of the survey of industrial users of thermal insulations, contact with consultants, and a brief review of economic analysis applied to thermal insulation, the following observations are made.

> Insulation of all heated surfaces—such as flanges, valves, and tanks—not presently insulated is economically justified for surfaces other than in waste streams.
>
> Availability of investment capital restricts many companies from upgrading insulation and other process improvements.
>
> Analysis of economic insulation thickness can lead to optimistic financial savings if the analysis uses unrealistic service life and thermal conductivity values. Achieving the desired service life emphasizes good design, installation, and maintenance procedures; realistic thermal property information emphasizes the importance of realistic data for service conditions.
>
> Many different economic criteria are used for making investment decisions in industrial companies. These criteria range from the simple payback period and return-on-investment to the discounted-rate-of-return based on discounted cash-flow analysis.
>
> Thermal insulation investment in new plants is justified to the economic thickness even with 30 to 50% per year discounted rates-of-return required.
>
> ECON-I and R-ECON manuals for calculating economic insulation thickness are generally not used by large industrial companies. Such companies use economic analysis programs that are more flexible, better understood, and hence more trusted. Also, the results from ECON-I and R-ECON often indicate such great insulation thicknesses as to be impractical because of physical confinement between equipment.

REFERENCES

(1) Malloy, J.F., *Thermal Insulation*, Chapter 2, pp. 89-126, Van Nostrand Reinhold Company, New York (1969).
(2) Bierman, H., Jr., and Smidt, S., *The Capital Budgeting Decision—Economic Analysis and Financing of Investment Projects*, 2nd ed., The MacMillan Company, New York (1966).
(3) *How to Determine Economic Thickness of Insulation*, National Insulation Manufacturers Association, New York, New York (1961).
(4) *ECON-I, How to Determine Economic Thickness of Thermal Insulation*, Thermal Insulation Manufacturers Association, 7 Kirby Plaza, Mt. Kisco, New York (1973).
(5) *R-ECON, A Method for Determining Economic Thickness of Add-On Thermal Insulation*, Thermal Insulation Manufacturers Association, 7 Kirby Plaza, Mt. Kisco, New York (1975).

ASSESSMENT
OF HEAT TRANSMISSION INFORMATION

The material in this chapter was excerpted from a report prepared by Oak Ridge National Laboratory (TID-27120).

INTRODUCTION

This chapter reviews some of the many facets of heat transmission through thermal insulations that have been found in open literature publications, company
brochures, and personal contacts. It is not a complete accounting of insulation
properties.

Most knowledge on insulations has been obtained by work near room temperature, with subsequent measurements and applications at higher and lower temperatures generally being outgrowths of studies at ambient temperatures. For
this reason insulations are discussed generally, without necessarily distinguishing
between building and industrial insulations. For instance, it is believed that
knowledge of the mechanisms of heat transmission operative in insulations and
subsequent control of the factors that influence these mechanisms form an integral part of any insulation application, be it building or industrial.

Clearly, advances in insulation technology must involve a circuit with information
feedback connecting the three areas — material development, subassembly tests,
and field evaluations. This assessment found that the documentation of knowledge decreases as one moves toward field evaluations, thus this review deals with
the material development area. Despite this shortcoming there is sufficient information and lack of information to warrant considerable discussion and concern
about insulation heat transmission.

Thermal insulations include a wide variety of materials that may be considered
mixtures of gases and solid bodies, which in large measure owe their insulating
value to the low thermal conductivity of the gas phase. As noted previously,
the insulation may be fibrous, granular, cellular, or multilayered and used as
evacuated or unevacuated systems.

They exhibit the range of thermal conductivity values k, from 10^{-5} to 10 W m^{-1} K^{-1} (7 x 10^{-5} to 7 Btu in hr^{-1} ft^{-2} °F^{-1}) as shown in Figure 11.1 taken from Pratt (1).

Figure 11.1: Range of Thermal Conductivities of Low-Conductivity Materials at or near Room Temperature (1)

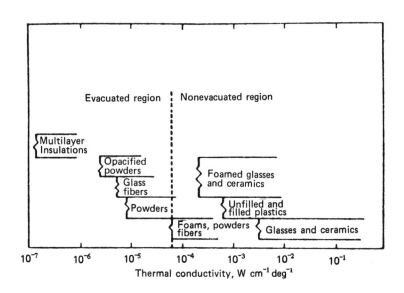

Source: TID 27120

Specific insulations show various thermal conductivity values and temperature dependences, as depicted in Figure 11.2, depending on the heat flow contribution of four basic mechanisms.

1.   Solid conduction through the materials forming the in-sulation, k_s;

2.   Radiation transfer through the voids and the components of the insulation, k_r;

3.   Gas conduction within the voids, k_c;

4.   Convective transfer in and through the voids, k_{cv}.

This chapter describes the role of the ASTM and industry in insulation, equipment used to measure insulation thermal conductivity, the information background on heat transmission depicted in Figures 11.1 and 11.2, the heat flow mechanisms, and factors that change the heat transmission behavior of insulations.

Figure 11.2: Relationships Between Thermal Conductivity and Temperature for
Typical Thermal Insulations and Other Low-Conductivity Materials

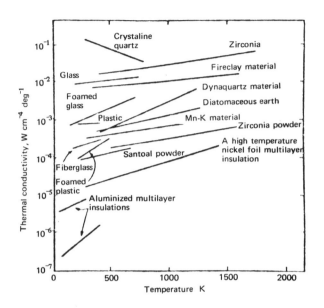

Source: TID 27120

Two further points of perspective are as follows. The thermal conductivity of
an insulation has an important bearing on the use of a given insulation; however,
this is just one of many properties that are needed for an insulation decision.
The term, thermal conductivity (k) has been used to describe the heat transmission value for an insulation; this term may or may not be appropriate, depending on the material, test conditions, and usage factors.

ASSESSMENT OF ASTM ACTIVITIES

The American Society for Testing and Materials (ASTM) serves as a focal point
for many insulation interests in the U.S. ASTM committees C-8 and C-16 are
particularly interested in property measurements on insulations. Committee
C-8, entitled Refractories, Glass, Carbon and Other Ceramic Materials, has primary interest in refractory materials, which may or may not be thermal insulations but are generally to be used above 650°C (1200°F), whereas committee
C-16, entitled Thermal and Cryogenic Insulation Materials, is concerned with
thermal insulations, generally for but not restricted to use below 650°C (1200°F).

Membership of these committees includes manufacturers, users, and measurers,
who are interested in standards and tentative specifications, methods of test,
recommended practices, definitions, and related materials, such as proposed meth-

ods. Parts 17 and 18 of the 1974 books of ASTM Standards (2) include the standards of C-8 and C-16, respectively. These committees hold business meetings semiannually to accomplish their assigned tasks, and they have made significant contributions to insulation technology. Actually, two subcommittees, C-8.04 and C-16.30, have responsibility for maintaining and updating the prescriptions for several methods of testing the heat transmission characteristics of insulation. The ASTM methods and responsibilities are listed below.

Subcommittee C-16.30

C 177-71	Thermal Conductivity of Materials by Means of the Guarded Hot Plate
C 518-70	Thermal Conductivity of Materials by Means of the Heat Flow Meter Method
C 236-66	Thermal Conductance and Transmittance of Built-Up Sections by Means of the Guarded Hot Box
C 335-69	Thermal Conductivity of Pipe Insulation
C 691-71	Thermal Transference of Nonhomogeneous Pipe Insulation at Temperatures Above Ambient
C 745-73	Heat Flux Through Evacuated Insulations Using a Guarded Flat Plate Boiloff Calorimeter

Subcommittee C-8.04

C 201-68	Thermal Conductivity of Refractories

Application of these methods has assisted in maintaining a degree of uniformity in insulation testing. However, it has been noted that these specifications represent a lowest common denominator of concerns, and can be misapplied and yield misleading results. Commensurate with this is the ASHRAE comment (3) about the Guarded Hot Plate method (ASTM designation C-177), "The accurate determination of fundamental conductivities and conductances requires considerable skill".

The ASTM Committee C-16 has sponsored four symposia that deal with thermal insulations over a broad temperature range (4)-(7). The initial paper in the proceedings (7) of 1973 ASTM Symposium was authored by the ASTM Subcommittee C-16.30 and entitled, "What Property Do We Measure?" This paper discusses the philosophy of the measurement of heat transfer properties of insulation and the manner of which certain heat transfer properties are used or possibly misused in describing thermal insulation performance.

The C-16.30 Subcommittee goal is to provide prescriptions of methods that measure applicable characteristic properties of insulations. Thus a material's thermal performance is determined at or near the conditions of actual use. In contrast, the practice in other countries is aimed at measuring theoretical properties of the material, which allows calculation of thermal performance.

In effect, the procedures used by other countries aim to obtain the thermal conductivity of the material, whereas the C-16.30 methods yield a thermal conductance, which—depending on the material, the test conditions, and test results—may or may not be the material's thermal conductivity. Understanding this is a most important consideration where heat transmission includes convective or radiative effects, which is the case for many insulations. The prescriptions described by C–16.30 include conversion factors for various thermal conductivity units, and they recommend use of SI units for k, W m^{-1} K^{-1}, and thermal conductance, W m^{-2} K^{-1}.

In contrast, most industrial brochures use English units for k, Btu in hr^{-1} ft^{-2} °F^{-1}, and Europeans use kcal m^{-1} hr^{-1} °C^{-1}. This diversity of units impedes efficient communication and increases the chance for error in calculations. The ATSM Standards emphasize that the methods are to be used for samples dried to a constant weight before testing. It should be recognized that the thermal conductivity obtained will not necessarily apply, without modification, to service conditions. In addition to the moisture question, other factors, such as mean temperatures that differ appreciably from the test, a density significantly different from the test sample, or for some materials, the effect of temperature difference, could be appreciable. Thus, the *user* should be extremely cautious.

In 1974 the ASTM C–16.30 subcommittee released results of an equipment and standards survey obtained by contacting known users of thermal conductivity equipment in the U.S. and abroad. The primary intent of the survey was to select reference materials of low k to be used as standards. While much valuable information was obtained, the much-needed primary intent remains elusive. While the ASTM methods are generally acceptable to industry, a wide variety of other k-measuring apparatuses have been used successfully to study insulating materials.

ASSESSMENT OF INFORMATION ON INSULATION

Since the subject of thermal insulation is so broad a tremendous amount of information is available. The quality of the information, however, is almost impossible to assess for the simple reason that there is such a wealth of data on such a variety of materials.

If, for example, the available data for a single type or class of industrial thermal insulation were plotted, the paper would be essentially black with points and curves. Therefore, each user will tend to pick a single source of information depending upon his own background and training. Typical resource materials might be the *Handbook of Chemistry and Physics* (8), Mantell's *Engineering Materials Handbook* (9), Mark's *Mechanical Engineers' Handbook* (10), Norton's *Refractories* (11), Perry's *Chemical Engineer's Handbook* (12) or the *ASHRAE Handbook of Fundamentals* (3).

Unfortunately, each of these sources is incomplete to a lesser or greater extent since they do not address all the pertinent properties of the materials such as strength, permeability, expansion, pH, combustibility, thermal shock resistance, etc., that might be especially important for a particular design or installation.

An example of the information that one typically finds in the literature (3)(13) for thermal conductivity are shown in Figure 11.3. As a general rule, much more may be found for k of insulations than for the other properties. The standard tests and specifications of the ASTM C–16 Committee on Thermal Insulations may be useful for defining, within rather broad limits, typical properties of a certain generic insulation.

Figure 11.3: Apparent Thermal Conductivity Versus Temperature for Various
Materials in Air at 1 atm (0.1 MPa) (13)

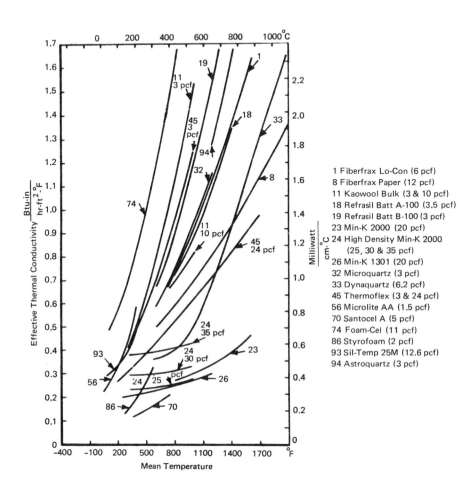

Source: TID 27120

Figure 11.4 shows some of the *k*-values specified by the ASTM and the corresponding density ranges. It must be remembered, however, that the ASTM specifications
are typically a "lowest common denominator," arrived at by committee concensus
and do not describe a specific trade-name insulation.

For example, the specification might call for maximum allowable density, a minimum strength, and a maximum conductivity. A product that meets the minimum
strength requirement will have a lower density and a lower conductivity than required by the specifications. For another product, for which greater strength is
required, the manufacturer will more closely approach the maximum density and

maximum conductivity, and will far surpass the minimum strength. Thus, the ASTM specifications are not of great value as a source of information if more than one manufacturer produces a given product. At the present time, the world's most complete compilation of thermophysical properties of materials is probably that of the Thermophysical Properties Research Center (TPRC—now known as Thermophysical and Electronic Properties Information Analysis Center) at Purdue University, which covers high-temperature solid materials (14).

Figure 11.4: Thermal Conductivities and Sample Densities According to the ASTM Specifications (2)

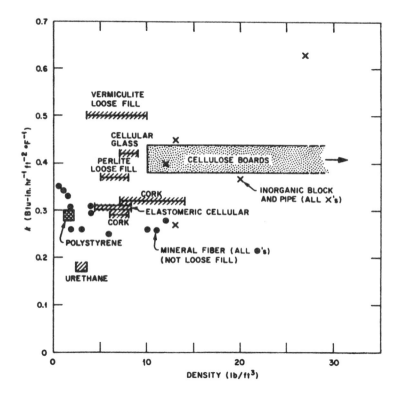

Source: TID 27120

The coverage of thermal insulations in this work, however, is weak, consisting of four spacecraft composite insulations, four organic foams, two inorganic foams, two inorganic powders, and the 1600 to 3100 IFB series. TPRC has issued a report (15) that describes polystyrene and polyvinyl chloride thermophysical properties, including those of the foams above 100 K (–279°F). A notable addition to the storehouse of knowledge on thermal insulations is a 1967 publication (13) prepared under NASA contract by A.D. Little, Inc., entitled *Thermal Insulation Sys-*

tems, A Survey. The focus is on the more exotic insulation systems such as multi-foils, but the report does include much useful information about all types of materials. The British-originated *The Insulation Handbook* (16) provides another useful compilation of properties and products. An extensive compilation of insulation properties was generated in 1962 under a USAF contract to Midwest Research Institute (17).

It should also be noted that for those materials used at reduced pressure, the variation of conductivity with pressure is often presented. The most significant observation about these compilations is that the data spread may range up to a factor of 8 for a given generic material. A valid reason for a large portion of this data spread is the material itself.

Over a period of time, a certain product will undergo significant alteration through changes in raw material and/or methods of manufacture. Thus, data for a glass fiber blanket actually represent a wide variety of samples from different manufacturers produced over a long time span. These factors, together with uncertainties in measurement techniques, contribute to the observed spread in the data.

A significant quantity of property information on thermal insulations may be found in the open literature. More than 36 papers dealing directly with insulations have been presented at the fourteen Thermal Conductivity Conferences held regularly since 1961. Publications arising from the regular meetings of the International Institute of Refrigeration also contain much information.

Not to be overlooked are the several Special Technical Publications of the ASTM and the whole host of technical journals such as those of the ASME, AIChE, ASHRAE, and American Ceramic Society, to name only a few. While much useful information is contained in all of these sources, it is difficult to assimilate. For example, in most instances, the materials tested are only poorly identified (e.g., from manufacturers A, B, and C) and are even more poorly characterized.

In addition, the experimental conditions are not suitably described with details such as surface conditions of critical apparatus components, temperature differences used, criteria of the experiment such as drift rates, and many other possible factors. The authors have encountered this same situation in much of the technical literature on thermophysical properties of all materials.

One source of information on insulations that should be discussed is the manufacturer. By use of either his own testing facilities or outside laboratories he studies his own products (and those of his competitors) and distributes this information as technical sales literature and brochures. In addition, he normally includes a disclaimer to the effect that property values listed are typical, and products as supplied are subject to normal manufacturing tolerances. Unfortunately, these normal manufacturing tolerances are almost never quantified, but indications are that the product variability may easily reach 25%. This is one of the most limiting factors in the entire assessment of insulation information.

The best source of information on thermal insulations that has been found is Malloy's book, *Thermal Insulation* (18). Following a thorough discussion of the properties of insulations, Malloy states the following.

Determining what insulation properties are important only indicates the information needed and provides the basis of investigation. The next question then becomes, how can these properties be measured and under what conditions should they be measured? Another question which follows is whether or not measurements made by laboratory test are truly indicative of the properties of the material in service. Does aging affect the properties? If so, how much and at what rate? What effect does temperature have on test results ? What effect does moisture content have on the properties?

Unfortunately, testing methods and studies of properties of material are not yet able to answer all these questions. Nor is there a laboratory available which is accepted as the authority on testing of properties of insulation materials. This statement does not imply that there are no standard test methods and laboratories to perform these standard test methods, but it does mean that the insulation industry lacks even the minimum requirements to provide reliable information. In many instances the manufacturers of insulation have developed their own test methods and can furnish some information regarding their materials. However, the test method used by one manufacturer is not always identical with that of another, so that the values of one cannot be compared with the results obtained by another. This makes direct comparison of material properties quite difficult. Another problem is that most of the mechanical properties are measured at atmospheric conditions and not at the temperatures to which the material is subjected in service. Inadequate information leads to poor design, the use of improper materials, and the installation failures to which the insulation industry has become accustomed.

In recent years progress has been made in correcting this confusion and presently there exists a number of American Society of Testing and Materials Standard Test Methods. However, even with standard test methods, sometimes it is necessary, because of the wide range in characteristics, to perform more than one test for a single property, as one test may be incapable of measuring all the characteristics for the wide variety of insulation materials. Thus, duplication of test methods for a single property of insulations does exist and the proper selection must be made for the insulation being tested.

After listing more than 40 ASTM test methods applicable to insulations, Malloy continues with the following.

As can be seen from the above, standard test methods do not exist for all the properties of insulation listed previously. One reason is that some of these properties are difficult to determine. In other cases test equipment is quite expensive, few manufacturers or laboratories have the required equipment, and an industry standard has never been devised.

It should be remembered that any laboratory test is a measure of a small sample. For certain characteristics, a small sample

will provide an accurate appraisal, whereas for other materials
the answer may only indicate the characteristics of that insulation. Other tests may attempt to predict the service life of insulation. Such service life tests are quite difficult to devise and
evaluate. Yet this type of test is essential, and as the demand
for it increases, eventually such tests will be devised by the industry.

It is believed that these statements present an accurate picture of the present
state of the art regarding the property information on insulations. Preceding the
presentation of the very extensive tabulation of properties of all types of insulations, Malloy cautions the reader thus:

Another factor which should be remembered in the use of
these tables is that a particular generic material may be made
by several manufacturers. As there are differences in the properties of a generic material produced by different manufacturers,
the properties are a mean of the values listed for the products
of the various manufacturers. Thus, for those products made by
several manufacturers, no single manufacturer's product will have
exactly the values as listed. It follows then, that these tables are
presented to serve as a *guide* for selection of the correct *generic
material* for an installation. However, the final choice of the particular manufactured product must be made by the individual by
comparing competitive products.

The tables presented by Malloy are probably the most complete of all presently
available, but significant gaps are obvious. Regarding these, he states the following.

Hopefully, this book will indicate the need for more complete information on properties of materials and that any
missing information will be made available by the manufacturers. When insulation consumers understand the need for
designing insulation systems on a scientific basis the manufacturers will quickly respond by providing the necessary technical information.

In summary, there is on the one hand, a wealth of information available on thermal insulation properties, but on the other hand, a severe shortage of really useful, high-quality data and analysis. Probably the most serious paucity is a clear
picture of the real product variability as supplied since this one factor has such
a profound effect on the worth of all property information.

For a number of years the previously mentioned heat transmission measurement
prescriptions and techniques have been applied to a variety of insulations to
measure the effect of variables on heat transmission. Numerous effects have been
identified, and mathematical descriptions of the phenomena have been suggested
to describe the various heat transport processes acting in the insulations.

One goal of such an approach is to obtain a useful understanding of the phenomena occurring in insulations and to alter or control the insulation characteristics
to reduce the heat transmission. Indeed this technological approach has yielded

improved insulations such as Min-K (Johns-Manville) and cellular materials, wherein the gas mean free path has been reduced, and superinsulations, wherein all conduction and convection components have been reduced by evacuation and the radiation component minimized by control of surface emittances.

Another goal is to predict the behavior of an insulation from a relatively few input parameters; however, because insulations involve such a variety of materials assembled in a variety of structures, this approach may be too simplistic. The alternative is to measure each insulation under the conditions of use, but this is time-consuming, expensive, fraught with errors, and possibly capable of only limited advances in insulation technology.

The theoretical studies on insulations have identified various mechanisms of heat transfer acting within the insulation. These mechanisms include; the solid components with their contact points, the gaseous conduction component and its characteristics, the radiative component and variables that alter it, and finally the gaseous convective component. Studies of these mechanisms are providing improved understanding of the behavior of insulations and, as indicated, often lead to new materials.

OBSERVATIONS FROM MEASURERS

There is a general feeling that the ASTM methods can be misapplied, are somewhat too lax, and lack enforcement. Careful and definitive measurements are made by only a small number of laboratories. One manufacturer has an insulation center to provide expert advice on insulation, as well as an institute to school contractors.

Opinions expressed about the overall quality of insulation property data available from the manufacturers range from complete satisfaction to total distrust. Many persons felt that the manufacturers had obtained much data in their own laboratories or from various user experiences, but had not released the information.

Some manufacturers and users claim that thermal conductivity tests are not of critical importance. However, this attitude may be fostered by the relatively high cost of these tests. Also, one must recognize that product variability and test conditions are such that a single test may be meaningless. Some sources acknowledged that some manufacturers had detailed information about their products' variability, inferred from property correlations or determined directly by conductivity tests. However, this information is not made available to the user, so he must make an educated guess of probable maximum and minimum values for use in his design. This approach will almost surely lead to conservative and non-cost-effective designs.

A problem voiced by both manufacturers and users was the inapplicability of the various ASTM test procedures to the situation in the real world. To be specific, the procedures state that the materials will be tested in a dry condition, a rare state in many insulation installations. However, the manufacturers correctly claim that this dry condition is the only state that can be routinely reproduced in the laboratory and that it would be futile to attempt to duplicate the many and varied conditions that might exist in the field.

Thus, the only guidance that the user has is the knowledge that the manufacturers' data are for dry conditions. His problem is then to estimate the degradation in conductivity caused by moisture or adsorbed gases and liquids. Careful experiments on this degradation have not been done for the obvious reason that they would be extremely difficult to perform and interpret. Therefore, there is an unresolved dilemma.

A second, specific complaint about the ASTM test procedures is that the smaller devices (most economical to construct and operate) allow tests on samples 25 mm (1 inch) or less in thickness, while practically all applications involve thicknesses greater than this. A thermal conductivity value (i.e., per unit thickness) is the result of the ASTM tests, while the user is more concerned with a thermal conductance (i.e., per total thickness).

A conductance may, in principle, be calculated from a conductivity only if convective or radiative conduction processes are not operable in the particular case. A few manufacturers acknowledge this problem and have equipment to test useful thicknesses of insulations, while others claim such tests would be prohibitively expensive because of the size of the apparatuses required.

Some sources believe that the suppliers of cryogenic superinsulations are the leaders in the area of competence. This may, in large measure, be attributed to the large expenditure of funds for aerospace research in recent years. This segment of the industry has used a systems approach to an insulation installation.

Thus, the design engineers for a cryogenic system are using data from their own company's laboratory plus knowledge gained from similar designs already executed. Since they are, in general, working toward a performance specification in the field, they are involved from beginning to end, and the customer is concerned only with the final result, not with the values used in the design. This is unique, and it would appear that this systems approach would be very advantageous to all users of thermal insulation, even for less critical applications.

Quite a different picture emerges at the opposite end of the temperature spectrum. The higher the temperature, the more ill-defined insulation properties are. This may be attributed to the difficulties encountered in making very-high-temperature measurements, to the absence of reliable reference materials, and to the lack of any firm theoretical basis for estimating or predicting property values.

Indeed, one refractory ceramic manufacturer scrapped its C 201 refractory conductivity tester because the time and effort involved in measurements was considered unwarranted. Considering the tremendous interest in thermal insulation that has developed in the past few years, it is an anomaly that the Refractories Research Center at Ohio State University made only five tests on thermal insulations [four ceramic fiber blankets, one 2600°F IFB (insulating fire brick capable of service up to 2600°F)] in 1976 for various clients.

This is not to say that the manufacturers are not concerned with high-temperature insulations, particularly the fiber blanket materials. Potential users certainly are, but they express strong doubts about the validity of the manufacturers' claims of the potential savings resulting from the application of these materials. One manufacturer has acknowledged the existence of these doubts and is presently constructing a special furnace with a 6 by 8 ft (1.8 by 2.4 m) instrumented test wall in order to make direct measurements of the thermal performance of various furnace wall constructions. It would seem that this approach to a systems tester is certainly a step in the right direction by the manufacturers and will yield much useful information. Another manufacturer has built a special mobile kiln to demonstrate the performance of ceramic fiber blankets.

Another indication of the level of interest in high-temperature insulations expressed by the manufacturers was their desire to conduct a round robin experiment on a fiber blanket material. This round robin would involve about eight laboratories, both manufacturers and private labs, and would be conducted on high-temperature guarded hot plates (C 177) and refractory testers (C 201). This comparison of methods and apparatuses is very timely and needed. A well-equipped unbiased laboratory is needed to coordinate this activity.

Germane to the discussion on ceramic fiber blanket insulation are comments from representatives of both C-8 and C-16 ASTM committees. The C-16 committee on thermal insulations, and particularly the C-16.30 subcommittee on measurements, has avoided the ceramic high-temperature insulations since this class of materials has been under the C-8 refractories committee.

These two are in conflict over the testing of ceramic fiber blankets and the operation of the C 201 tester. This needs to be resolved because these materials are becoming increasingly important, and the present situation may hinder development of needed standards and specifications.

The region extending from room temperature down to liquefied-natural-gas temperatures is a particularly troublesome one to the insulation user, more than to the property measurer. For this temperature range, organic foams, bulk loose-fill inorganics, or rigid inorganic forms of insulation are applicable. Herein lies the dilemma for the insulation buyer since he is confronted with often conflicting information about the competing products. Comments from the user indicated that a major concern was not over the initial properties of the particular insulation, but rather was in the change in properties with time in service.

The problem arises primarily from water vapor condensation and migration within the insulation. In many low-temperature applications, ice will form within the insulation, mechanically destroying it. The insulation manufacturers are certainly aware of this problem, but believe their responsibility is to provide insulation and barriers and to recommend good procedures for its installation. They realize that the properties of their material will be significantly altered (always degraded) by the condensed gases, but cannot qualify the magnitude of the effect because of the many variables.

An insulation consulting engineer claims that some of the problems might be resolved if more realistic diffusion rates or permeabilities were quoted for products. He found moisture permeabilities 100 times greater than the stated values and attributed this difference to unrealistic test conditions.

In addition to property changes due to moisture, some materials—specifically the organic foams—double in thermal conductivity because of replacement of the foaming gas by air over a period of time. This is a natural diffusion process and is strongly temperature dependent. Users stated that this had occurred in their systems at a rate much faster than anticipated and required the installation of additional insulation to the system.

There is some doubt about the manner in which the manufacturers' property data was acquired. Manufacturers quote a *thermal conductivity* obtained using a temperature drop ($\triangle T$) through the sample, which is quite proper under the ASTM procedures. Often the user is concerned with the *thermal conductance* of the insulation under the operating conditions, which typically involves a much larger $\triangle T$. Thus, radiative heat transfer might become very significant, and the conductance could be much higher than indicated by the measured conductivity if the material is semitransparent at the wavelengths involved.

This shows variability of applications and the importance of the details of testing procedures and the problems that might arise if the manufacturer and the user do not fully understand all the implications in the numbers. The suggestion was made that the manufacturers should test under service conditions exclusively and so state in their literature. The manufacturers counter by stating that this information is available currently.

Much of the preceding discussion is also relevant to the intermediate temperature types of thermal insulations, although the variety of products becomes more limited as the temperature increases. Users encounter the moisture problem, but without the complication of ice formation. It is also in the temperature range that users are often more concerned with properties other than thermal conductivity.

The comment was often made that the last property asked for was k. Within the compromises of an installation this may be reasonable since the k values for most insulations of a given temperature class are nearly the same. Manufacturers provide a product to fit a set of needs, with a k that is more of a compromise with other properties than a target.

Manufacturers and users generally feel that the thermal conductivity data for the insulations in the room temperature to intermediate temperature range are of reasonable quality. In this temperature range, however, one must recognize the possibility of radiative heat transfer and, as discussed earlier, must determine if the reported conductivities are realistic for the particular situation at hand.

The preceding indicates that, with the notable exception of the cryogenic segment, the whole insulation picture is clouded by fragmentation. That is to say, the person that makes the measurements is not really involved with their application; the designer is probably not familiar with the details of the measurements; the distributor fills orders; the installer follows his own practice; the customer gets an insulation system that may work; and if it doesn't, to whom does he turn? Responsibility is divided and there is no logical overall control.

One of the critical failings is that no one really follows the entire procedure to the logical conclusion of the job (i.e., a detailed testing and analysis of the insulation system as installed). Thus, the loop is not closed, and without this closure, the validity or quality of the data, the design, and the installation cannot be assessed.

Although it may be expensive, adoption of a performance specification approach for an insulation system, as is done in the cryogenics area, may be a fruitful area to pursue. Now may be the time to examine the merits of a move in this direction.

CONCLUSIONS

1. Advances in insulation technology are being limited by the lack of open-literature publications and information feedback in three interrelated (but now isolated) areas: materials development, subassembly testing, and in-field evaluation.

2. ASTM activities have been invaluable and vital to insulation technology and have resulted in: a degree of uniformity in prescriptions for testing insulations, four symposia on insulations, and a survey of equipment for measuring thermal conductivity.

3. Thermal conductivity or conductance tests on insulations are difficult to perform and consequently they are often expensive, are fraught with incipient errors, and yield questionable results. Acceptable materials for thermal conductivity standards are virtually nonexistant.

4. Most of the available apparatuses, which are based on ASTM test method prescriptions, are useful for tests on product thicknesses that are much less than are actually used in installations. Current practice is to extrapolate thin sample results to thick sample applications, although justification of this extrapolation may not be valid.

5. The ASTM maintains test prescriptions on at least seven methods of measuring heat transmission. Numerous other methods have been developed to study heat conduction in insulation. Proper use of all techniques requires close attention to experimental detail to obtain valid results. Often this close attention is missing, so the results are questionable. There is no panacea for this problem. Few of the scattered published works provide data comparisons with other work, and often the claimed checks on accuracy, reproducibility, and repeatability are inadequate and optimistic. There appears to be no adequate enforcement of policies to remove this difficulty, and this void produces the least common denominator approach, which denies any significant advance in methodology on these complex materials.

6. There appears to be no complete single source of valid heat transmission data on thermal insulations. For instance, data are not available on thermal insulation product variability, which may reach 25%, even though this is crucial to the performance and design of many installations.

7. The heat flow through insulation has been identified as dependent on the contribution and interaction of four basic mechanisms: solid conduction, gas conduction in voids, convective transfer in and through voids, and radiation transfer through voids and components of insulation. Studies directed at obtaining a better understanding of the roles of these mechanisms have often led to improved products and installations by identifying and controlling certain material characteristics. Such studies have been conducted on the various generic insulation types including fibrous, cellular, granular, bulk, and evacuated insulations.

8. Relatively few studies have been conducted on in-service variables and their influence on heat transmission in insulations. The effect of moisture in insulations is known to be quite detrimental. However, because the ASTM guides call for tests in dry conditions, definitive studies on moisture effects are not being undertaken.

9. Numerous comments and observations from those involved in heat transmission in the thermal insulations field indicated a spectrum of views. Misapplication of ASTM methods, scrapped equipment, unused facilities, outmoded facilities, lack of reference materials, undefined specifications, limited system testers, few practical studies, and lack of turn-key systems are some of the key phrases used to describe the current arena of thermal insulations.

REFERENCES

(1) Pratt, A.W. Chapter 6, "Heat Transmission in Low Conductivity Materials", pp. 301-405 in *Thermal Conductivity*, Vol. 1, ed., R.P. Tye, Academic Press, London (1969).
(2) *1974 Book of ASTM Standards with Related Material*, Part 17 (C-8) and Part 18 (C-16), American Society for Testing and Materials, Philadelphia.
(3) *ASHRAE Handbook of Fundamentals*, Chap. 18, p. 242, American Society of Heating, Refrigeration, and Air-Conditioning Engineers, New York (1968).
(4) *Symposium on Thermal Insulating Materials*, Am. Soc. Test. Mater. Spec. Tech. Publ. 119, American Society for Testing and Materials, Philadelphia (1952).
(5) *Symposium on Thermal Conductivity Measurements and Applications of Thermal Insulations*, Am. Soc. Test. Mater. Spec. Tech. Publ. 217, *American Society for Testing* and and Materials, Philadelphia (1957).
(6) *Thermal Conductivity Measurements of Insulating Materials at Cryogenic Temperatures*, Am. Soc. Test. Mater. Spec. Tech. Publ. 411, American Society for Testing and Materials, Philadelphia (1967).
(7) *Heat Transmission Measurements in Thermal Insulations*, Am. Soc. Test. Mater. Spec. Tech. Publ. 544, American Society for Testing and Materials, Philadelphia (1974).
(8) *Handbook of Chemistry and Physics*, 46th ed., The Chemical Rubber Company, Cleveland, Ohio (1965-66).
(9) *Engineering Materials Handbook*, 1st ed., ed. by C.L. Mantell, McGraw-Hill, New York (1958).
(10) *Mechanical Engineers' Handbook*, 5th ed., ed. by L.S. Marks, McGraw-Hill, New York (1951).
(11) Norton, F.H., *Refractories*, 3rd ed., McGraw-Hill, New York (1949).
(12) *Chemical Engineers' Handbook*, 3rd ed., ed. by J.H. Perry, McGraw-Hill (1950).
(13) Glaser, P.E., et al, *Thermal Insulation Systems, A Survey,* NASA Spec. Publ. 5027, National Aeronautics and Space Administration, Washington, D.C. (1967).
(14) *Thermophysical Properties of High Temperature Solid Materials*, Vol. 3, ed. by Y.S. Touloukian, The Macmillan Company, New York (1967).
(15) Ho, C.Y., et al, *Thermophysical Properties of Polystyrene and Poly (Vinyl Chloride)*, CINDAS-TRPC Report 38 (August 1975).
(16) *The Insulation Handbook 1973*, Lomax, Wilmoth and Company, London.
(17) *Thermophysical Properties of Thermal Insulating Materials*, ASD-TDR-62-215, Wright-Patterson Air Force Base, Ohio (July 1962).
(18) Malloy, J.F., *Thermal Insulation*, Van Nostrand Reinhold Company, New York (1969).

SOURCES UTILIZED

The following reports were used in the preparation of this book:

BNL-50862

An Assessment of Thermal Insulation Materials and Systems for Building Applications, prepared by Brookhaven National Laboratory with the assistance of Dynatech R/D Corporation, for U.S. Department of Energy, June 1978.

DOE/CS-0051

Material Criteria and Installation Practices for the Retrofit Application of Insulation and Other Weatherization Materials, a technical report by the Division of Buildings and Community Systems, Office of Conservation and Solar Applications of the Department of Energy, November 1978.

PB-259 937

ETI—Economic Thickness for Industrial Insulation, prepared by York Research Corp., for the Federal Energy Administration, August 1976.

TID-27120

R.G. Donnelly, V.J. Tennery, D.L. McElroy, T.G. Godfrey, and J.O. Kolb, Industrial Thermal Insulation—An Assessment, prepared by Oak Ridge National Laboratory, for the Energy Research and Development Administration, August 1976.

HOW TO SAVE ENERGY AND CUT COSTS IN EXISTING INDUSTRIAL AND COMMERCIAL BUILDINGS 1976

An Energy Conservation Manual

by Fred S. Dubin, Harold L. Mindell and Selwyn Bloome

Energy Technology Review No. 10

This manual offers guidelines for an organized approach toward conserving energy through more efficient utilization and the concomitant reduction of losses and waste.

The current tight supply of fuels and energy is unprecedented in the U.S.A. and other countries, and this situation is expected to continue for many years. Never before has there been as pressing a need for the efficient use of fuels and energy in all forms.

Most of the energy savings will result from planned systematic identification of, and action on, conservation opportunities.

Part I of this manual is directed primarily to owners, occupants, and operators of buildings. It identifies a wide range of opportunities and options to save energy and operating costs through proper operation and maintenance. It also includes minor modifications to the building and mechanical and electrical systems which can be carried out promptly with little, if any, investment costs.

Part II is intended for engineers, architects, and skilled building operators who are responsible for analyzing, devising, and implementing comprehensive energy conservation programs. Such programs involve additional and more complex measures than those in **Part I**. The investment is usually recovered through demonstrably lower operating expenses and much greater energy savings.

A partial and much condensed table of contents follows here:

Much of the technology required to achieve energy savings is already available. Current research is providing refinements and evaluating new techniques that can help to curb the waste inherent in yesteryear's designs. The principal need is to get the available technology, described here, into widespread use.

ISBN 0-8155-0638-4

725 pages

SOLAR HEATING
AND COOLING 1977
Recent Advances

by J. K. Paul

Energy Technology Review No. 16

The technology for solar energy utilization is becoming increasingly available, as indicated by the large number of patents issued in the past several years. Recent developments encompass a number of areas. The emphasis in this book is on low temperature (to +90°C) solar collector construction and heating and cooling systems which use these low temperature collectors. The material discussed here is based on 175 U.S. patents, issued since 1970, which illustrate 157 processes.

In its simplest form a collector consists of a sheet of glass or other transparent material situated above a flat plate so constructed that it acts as a black body to absorb heat. The sun's rays pass through the glass and are trapped in the space between cover and plate. The heat may then be utilized by passing a fluid through a conduit system located between the cover and absorber plate; the heated fluid subsequently being used to heat a home, water supply, or swimming pool, or even run a heat pump for cooling.

Focusing collectors use curved or combinations of flat devices to reflect solar rays onto an absorber surface to achieve greater concentration of energy (higher temperatures).

Information has been included describing suitable coatings used to improve absorption properties and detailing a number of devices which employ liquids, crushed rock or other media for the storage of absorbed energy when the weather is cloudy or hazy, or at night. A partial table of contents follows here. Numbers of processes are in parentheses.

ISBN 0-8155-0674-0

485 pages

SOLAR ENERGY
FOR HEATING AND COOLING
OF BUILDINGS 1975

by Arthur R. Patton

Energy Technology Review No. 7

Solar energy can be used for indirect heating purposes in many ways. The information in this book has been limited to so-called low temperature solar thermal processes. Designs requiring photocells or other thermoelectric generators and lenses or reflecting mirrors plus tracking equipment have been excluded.

Low temperatures are the easiest to obtain, and the necessary collectors are fairly simple in construction. A black surface is used to absorb the sun's rays, this surface is usually covered with glass and the collector is insulated on the back and sides against heat loss. Water or some other heat transfer fluid is passed through the collector and can reach temperatures from 60°C (140°F) to about 95°C (203°F). The thermal energy is then stored in a heat storage system (perhaps based on the latent heat of fusion of selected salts). Coupled to the heat storage system are heating loops to furnish heat by convection and to operate an air conditioning system. In most temperate zones an auxiliary heater, operated with conventional fuels, must also be connected.

Large scale applications designed for schools and similar building are beginning to appear or are in the planning stage. This book describes in detail several large scale feasibility studies with designs suitable for institutions and industrial plants.

Descriptions are based on studies conducted by industrial or engineering firms or university research teams under the auspices of various government agencies. A partial and condensed table of contents follows here.

ISBN 0-8155-0579-5

328 pages

UTILIZATION OF WASTE HEAT
FROM POWER PLANTS 1974

by David Rimberg

Pollution Technology Review No. 14

Energy Technology Review No. 3

Present-day steam-driven turbo-electric power plants in the United States discharge as waste heat an amount of energy roughly equivalent to twice their total electricity-generating capacity. This energy is most difficult to utilize, because it is degraded in temperature. The effluent water is warm, but it is far from being near the boiling point. It constitutes a necessary, but unwanted, by-product of the energy conversion process for generating electricity.

The growing quantities of waste heat discharged, and the increasing ecological anxieties about undesirable growths, energy utilization and thermal discharge problems, have stimulated an examination of methods for productively using energy presently wasted on the environment.

Part I of this book discusses the reasons for present-day ineffective utilization, but It also shows ways and means to promote more efficient energy usage.

Part II assesses the cause, magnitude, and possible effects of heat discharges into water from steam electric power plants and related condenser cooling systems. Also contained in this section is a discussion of the thermodynamics of the electric power generation cycle detailing the reasons for the "inefficiencies" in by-product heat generation.

Ultimately all the accessible waste energy appears as low temperature heat, and the term "waste heat utilization" refers to the performance of useful functions with this heat before it is discharged into the environment. Within the framework of this book, the minimum temperature of the effluent water considered for subsequent use is about 38°C or 100°F. These uses are discussed in **Part III**: Food production in agriculture, hydroponics and aquaculture (fish farming) and the use of heat in wastewater treatment.

In the past the dissipation of waste heat was accomplished by wet and dry cooling towers and cooling ponds, lakes and streams. These methods are now being challenged by some sectors of society, and industry is being forced to consider their environmental impact. In this regard **Part IV** discusses the research needs necessary to equate the complicated interactions of the physical, engineering, biological, and social aspects of this waste heat problem.

This Pollution Technology Review is based on studies conducted by industrial and engineering firms or university research teams under the auspices of various governmental agencies, e.g. The National Water Commission, Oak Ridge National Laboratory Federal Water Pollution Control Administration, Office of Water Resources Research, Environmental Protection Agency, National Science Foundation, and the Department of Commerce.

A partial and condensed table of contents follows here:

ISBN 0-8155-0555-8

175 pages

FUEL CELLS
FOR PUBLIC UTILITY
AND INDUSTRIAL POWER
1977

Edited by Robert Noyes

Energy Technology Review No. 18

Fuel cells are generators of electricity containing no moving parts except small extraneous pumps for the movement of fuel and oxidant into the cell and the products of oxidation out of the cell.

Public utilities and industrial consumers require high voltage, three-phase alternating current. In this application fuel cells must compete with turbine-driven generators which provide such current. While the output of a fuel cell is low voltage DC power, cells may be connected in various series and parallel arrangements to give whatever voltage is desired, but mechanical rotary converters or delicate electronic inverters must then provide conversion to AC (60 cycles for each phase in the USA).

The advantages of a fuel cell system over turbine-driven generators lie in greater efficiency at full load which even increases as the load diminishes, so that inefficient peaking generators are not needed. There are considerable pollution control advantages to be gained as well. Because the various suitable fuels react electrochemically rather than by burning in air, no nitrogen oxides are formed. For the same reason, emissions of unburned or partly burned gaseous and particulate products are practically nil.

Fuel cell installations using inverters have a long life with relatively little maintenance. It is now entirely feasible to have small, completely unattended fuel cell power plants using waste fuels on location (such as hydrogen from chlorine production) to produce convenient power.

This book, based on information derived from U.S. government-contracted studies, contains considerable practical down-to-earth technical information relating to fuel cells for power plants. A partial and condensed table of contents follows.

1. INTRODUCTION
Advantages of Fuel Cells
Application of Fuel Cells
Choice of Fuels

2. TYPES OF FUEL CELLS & SPECIAL APPLICATIONS
Space
Naval Propulsion
Electric Vehicles
Communication Systems
Commercial Systems
Power Plants

System Characteristics
Near Future Systems

3. ASSESSMENT OF FUELS AND POWER GENERATION
Preferred Fuels
Secondary Fuels
Gasification Technology
On-Site Alternatives
Modular Approach
Central Fuel Conversion
Synthesis Gas Treatment
Coal Gasification Systems
Heavy Oil Partial Oxidation

4. WESTINGHOUSE STUDY
Acid Fuel Cells
Alkaline Fuel Cells
Molten Carbonate Fuel Cells
Zirconia Fuel Cells
Solid Electrolyte System
Fuel Processing for Low
   Temperature Fuel Cell Power Plants
Steam-Methane Reformer
Shift Converter
Oxygen Plants for
   Fuel Cell Power Systems

5. GENERAL ELECTRIC STUDY
Low Temperature Fuel Cells
Solid Polymer Electrolyte (SPE)
Phosphoric Acid Cells
High Temperature Fuel Cells
Design & Cost Bases
Current Inversion (Costs & Losses)

6. FUEL CELL— POWER PLANT EVALUATION
General Electric and
   Westinghouse Studies
Significant Trends
Detailed Evaluations

7. MARKETING CONSIDERATIONS
Summary
Historical Overview
Operating Characteristics
Response to Variable Loads
Siting and Installation
Anticipated Regional Fuel Prices
   and Annual Generating Costs

8. PROPRIETARY PROCESSES
About 500 Abstracts of recent U.S. Patents
   on Fuel Cells, Fuel Cell Materials
   and Related Subjects. Organized
   by Companies.

ISBN 0-8155-0676-7

325 pages

COAL ASH UTILIZATION
FLY ASH, BOTTOM ASH AND SLAG
1978

Edited by S. Torrey

Pollution Technology Review No. 48

Ash is a waste product left after the burning of many combustible substances, and fly ash is the accepted term for the finely divided residue that results from the combustion of ground coal. It is easily disseminated by flue gases, unless checked and collected by suitable devices.

Fly ash particles are primarily composed of silica and alumina. Secondary ingredients are carbon and oxides of iron, calcium, magnesium and sulfur.

Energy and environmental considerations over the coming years point to greater use of coal: As more utilities are forced to shift from gas and oil to coal as a source for fuel, the available quantities of fly ash will increase. This book shows how millions of tons of fly ash and other ashes can be disposed of annually in an environmentally acceptable manner. There is, for instance, a well-developed technology using lime + fly ash aggregates in pavements, buildings and bridge construction. Since these aggregates generally require only 2 to 5 percent lime, the expenditure of money and energy for these mixes is very low, thus making an attractive case for the expanding use of lime + fly ash aggregates.

The book is based mostly on federally funded studies as listed in the bibliography at the end of the volume. A partial and condensed table of contents follows here.

ISBN 0-8155-0722-4

370 pages

THERMAL ENERGY
FROM THE SEA 1975

by Arthur W. Hagen

Energy Technology Review No. 8
Ocean Technology Review No. 5

Recent advances in heat transfer research and thermal power plant design suggest that sea thermal power can now be made competitive with more conventional generating methods.

Utilization of ocean thermal gradient systems appears relatively attractive from many points of view. The ocean acts as a large solar energy heat reservoir which reduces energy storage requirements and permits the system to be operated the year round, 24 hours per day.

The purpose of this book is to provide a condensed data base to aid in proof-of-concept experiments and continued R & D to prove the technical feasibility and economic viability of generating either electricity or hydrogen by harnessing the temperature gradients in the sea.

The optimum locations for such generating plants appear to be in latitudes from 23°N to 23°S. Transmission and storage problems may limit the amount of power delivered by ocean-based plants. Construction of power production facilities must consider design problems associated with the hazards of marine environment.

Reports of proposed solutions to these problems are also contained in this book which is based on government-sponsored studies by engineering firms and university research teams. A partial and condensed table of contents follows:

ISBN 0-8155-0597-3

150 pages

GEOTHERMAL ENERGY

Recent Developments

1978

Edited by M.J. Collie

Energy Technology Review No. 32

The heat underneath the Earth's crust seems a highly desirable energy source alternative to oil and gas. The Earth does not give up her energy readily, still geothermal energy activities are progressing on a broad front worldwide wherever possible.

Operational fields are increasing in numbers, and new and improved technologies, especially new turbine designs and deep-well pumps, coupled with heat exchangers at the surface, are leading to much greater operating efficiency.

Yet, as a rule, geothermal power stations, with a few exceptions, are inefficient in converting thermal energy to electric energy, because of the overall low temperatures. Also the highly mineral-laden waters in the geothermal reservoirs increase operating expenses through scale formation, thus complicating the process of extracting the heat.

Geothermal energy is not expected to play a large overall role in the U.S. But it can have regional significance, especially in the West. For developing countries, however, geothermal energy, especially in volcanic regions, may be the only economic form. Electricity systems there are too small to justify nuclear power stations, but the comparatively small size of geothermal power stations fits the scale of electricity supply systems in these countries.

This energy Technology Review is based on studies conducted under the auspices of various government agencies and the last chapter constitutes a survey of the recent patent literature.

ISBN 0-8155-0727-5

445 pages